Effective Troubleshooting with EVM & Scope

By Ben Gaddis

TAB BOOKS
Blue Ridge Summit, Pa. 17214

FIRST EDITION

FIRST PRINTING—SEPTEMBER 1974

Printed in the United States
of America

Hardbound Edition: International Standard Book No. 0-8306-4730-9

Paperbound Edition: International Standard Book No. 0-8306-3730- 3

Library of Congress Card Number: 74-81738

Cover photo courtesy Simpson Electric Company.

Preface

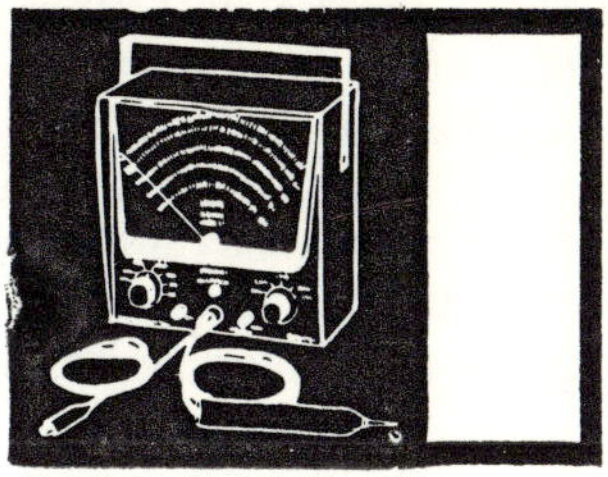

The meter and scope are the two most important pieces of test equipment a technician uses. With these instruments, the overall characteristics of electronic equipment may be observed, individual stages may be checked for proper operation, and even individual components may be tested. A large assortment of test gear is used in the average service shop, but most of this equipment is used in conjunction with the meter and scope. No other two instruments are so versatile and necessary.

If you are going to be successful in troubleshooting, you must know how to use these two basic pieces of test equipment. The purpose of this book is to show you how to use them to the best advantage. If you have a basic knowledge of electronics—maybe you are a hobbyist or a student—then this book is meant especially for you. It will also be useful as a bench reference for more experienced technicians.

The first chapter of this book explains some of the basic principles common to all meters. This is followed by a chapter pointing out the differences between electrical meters and the more sophisticated electronic meters. Then, the techniques of troubleshooting with the electronic voltmeter are explained.

Succeeding chapters examine the basics of waveform observation and oscilloscope operation. After this, we look at two specific oscilloscopes to explain how these instruments are operated—first a typical low-priced, wide-band service scope, then a more complex triggered-sweep scope. For the color television technician, basic vectorscopes are discussed next.

Two chapters are devoted to showing how the scope is used in actual practice. One looks at some of the more general uses of scopes, while the next covers television servicing applications. Emphasis in both of these chapters is on scope operation. The final chapter of the book gives some of the basic tests you can make on a scope to insure that it is operating properly—a very helpful chapter if you are in the market for a used scope.

I would like to thank all the manufacturers of test equipment who helped in supplying information for this book. Special thanks goes to the following companies who have supplied photographs, technical manuals, and advice: B&K Div. of Dynascan; EICO Electronic Instrument Co., Inc.; Heath Company; and Sencore Electronic Test Instruments.

Ben Gaddis

Contents

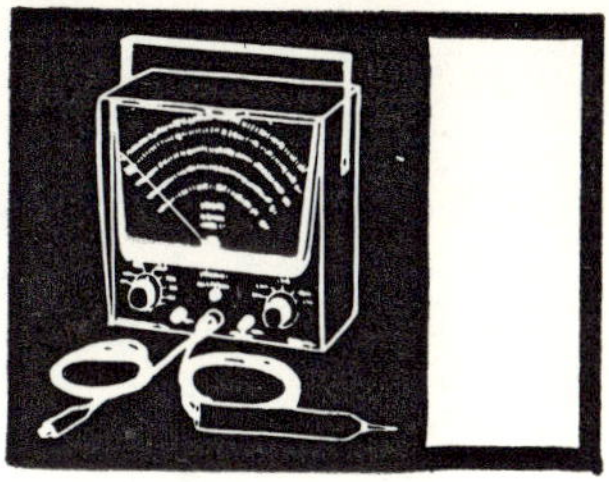

Basic Electronic Measurements

1

A troubleshooter is one who locates and eliminates sources of trouble. To locate any trouble, the troubleshooter must do a considerable amount of observing. The electronic technician must observe the basic operation of the device in question in various operating situations, continuing these observations until the fault can be isolated to some circuit. At this point, he may shift the scope of his test procedures and do some more observing, attempting to pinpoint the trouble to a specific component.

To observe the electronic circuit, the troubleshooter may use a wide variety of test equipment. Some of the tests will be made in what I call an "active" condition. Under this condition, the equipment under test is energized and supplied with the necessary inputs to cause the circuit to operate. Voltage, current, and frequency are the characteristics most often observed in this situation. The behavior of the circuit is judged from test equipment which monitors these three basic functions.

The other condition under which electronic circuits can be tested is the "cold circuit" situation. Here, the power is removed and the circuits tested for qualities that will affect operation when power is once again applied. At times it will be necessary to remove individual components from the circuit for individual testing. The most often observed characteristics under these conditions are resistance, inductance, capacitance, and tube or semiconductor characteristics.

Although comprehensive testing such as this requires a large selection of test equipment, there are two pieces of test gear that you will use more often than any others—the meter and the oscilloscope, or "scope." Knowing the characteristics and operating techniques of these two pieces of equipment is one of the most valuable troubleshooting aids you can have.

Like the circuits tested, the meter and scope have undergone considerable change since I replaced my first fuse some 20 years ago. Then, the average technician considered himself lucky if his meter had an input sensitivity of 20,000 ohms per volt. Now, portable, lightweight, versatile meters

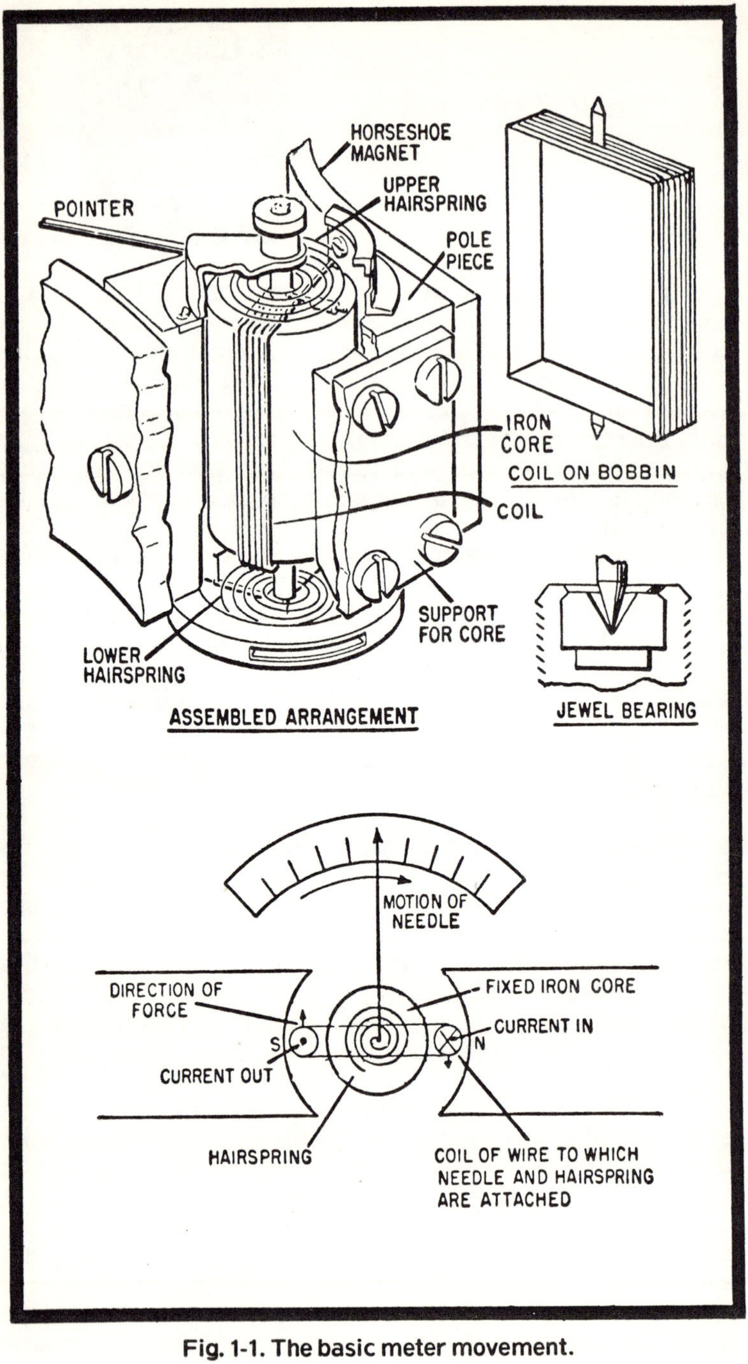

Fig. 1-1. The basic meter movement.

with **megohms** of sensitivity are within the price range of any technician. In fact, this new equipment is a "must" for servicing today's electronic circuits.

Since the majority of the observations made during troubleshooting will be concerned with resistance, voltage, and current, let's look at the principles of these measurements.

ELECTRICAL METERS

The most fundamental of all electronic measuring devices is the meter. There is a wide variety of ohmmeters, ammeters, and voltmeters. All use some sort of **meter movement** to indicate a quantity of measure.

A basic meter movement consists of a very fine coil of wire held by a shaft between the poles of a permanent magnet, as shown in Fig. 1-1. The shaft is mounted by means of jewel bearings that reduce friction, making the coil assembly easy to rotate. Small hairsprings fastened between the shaft and its support hold the coil in a position that causes the pointer attached to the shaft to point to the zero position on the meter scale. When a dc voltage is applied across the leads of the coil, current flows, establishing a magnetic field about the coil. The magnetic field reacts with the field of the magnet, resulting in a "motor" action that causes the coil to move. The movement of the coil rotates the shaft and its attached pointer.

Basically, any electrical meter is a current-operated device. The amount of current flowing in the coil determines the distance the pointer is displaced. Depending on the intended use of the meter, external circuitry may be used with the movement to cause the meter to indicate the value of various electrical quantities. The scale is **calibrated**, or marked, to read whatever quantity the meter is to measure.

The Ammeter

To measure current, the meter must be physically inserted in the current path so that the current to be measured passes through the meter. The meter is then **in series with** the circuit. The meter movement just discussed could be used in just this way to measure a small direct current. In order to measure larger currents, resistors are connected across the coil to bypass a portion of the circuit current around the meter movement. Resistors connected in that manner are called **shunt** resistors. The shunt carries the part of the current in excess of that usable by the meter. Using shunt resistors and a switching arrangement such as that shown in Fig. 1-2, the meter may be built to measure almost any desired value of direct current.

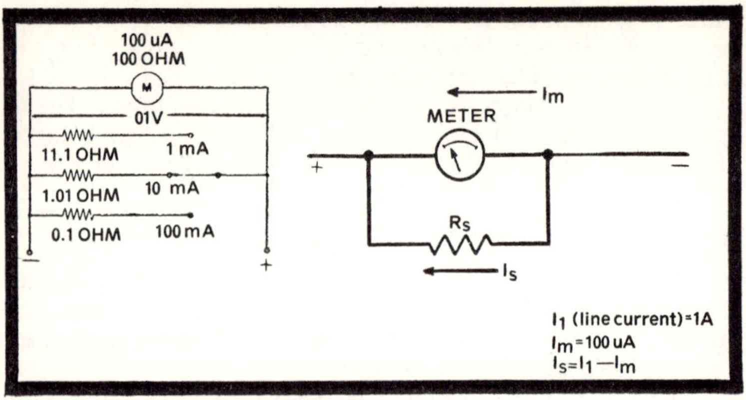

Fig. 1-2. The milliammeter.

For example, suppose you desire to extend the range of a basic dc meter in which the meter movement has a resistance of 100 ohms and a current capacity of 100 uA (microamperes) for full-scale deflection. Further assume that the current to be measured is less than 1A.

Since the meter deflects full scale when the current through the 100-ohm coil reaches 100 uA, the voltage drop across the coil that causes full-scale deflection is given by

$$E = IR$$

$$= 0.0001 \times 100 = 0.01V$$

The shunt resistor and the coil are in parallel; therefore, the same voltage drop, 0.01V, is also present across the shunt. The maximum current through the shunt is the difference between the current necessary for full-scale deflection and the circuit current to be measured. In this case, then, with a 100 uA meter, the shunt resistor must pass 0.9999A for the meter to read 1A at full-scale deflection. This can be computed as follows:

$$I_s = I_l - I_m$$

where I_s = shunt current
I_m = meter current required for full-scale deflection
I_l = maximum line current to be measured

Hence $I_s \ 1 - 0.0001 = 0.9999A$

The value of the shunt resistor is

$$R = E / I$$

where **E** is the voltage drop across the coil and **I** is the shunt current. In this case

$$R = 0.01 / 0.9999 = \text{about } 0.01 \text{ ohm}$$

The current distribution is shown in Fig. 1-2B. The current through the meter, I_m, is 100 uA for full-scale deflection. Line current I_l, in this case, is 1A. Therefore, 0.9999A must be shunted around the meter coil. Figure 1-2A shows how one meter movement may be modified to read four different ranges of current. Why four different ranges? Suppose you have a meter movement that has only one 1-ampere scale. It would probably have a scale with 10 major graduations, as shown in Fig. 1-3A, with each graduation having 5 subdivisions. If 1A causes full-scale deflection, each major division is equal to 0.1A, or 100 mA, and each small graduation is equivalent to 0.02A, or 20 mA. With the pointer at the position shown in A, the technician's "calibrated eyeball" can read 0.05A. That may seem pretty close. But look back at Fig. 1-2A. Let's assume that this is our meter and we have the four scales indicated. If the switch is placed in the **100 mA** position and the current just discussed is measured, the pointer will move to the position shown in Fig. 1-3B, if in fact the current was exactly 0.05A as measured originally. But by having the full-scale range decreased to 100 mA full scale, each major division now represents 10 mA, or 0.01A, and each small division represents one-fifth of 10 mA—2mA, or 0.002A. Note how that compares to each small division being equal to 0.02A in the previous example. Clearly, increased accuracy is the reason for the smaller scales. If in the example just explained, circuit current had been 0.051A, it is doubtful that even the best "calibrated eyeball" could have picked up the

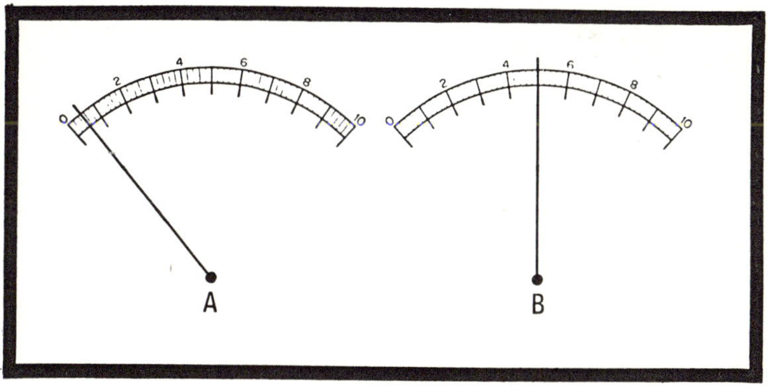

Fig. 1-3. The meter scale.

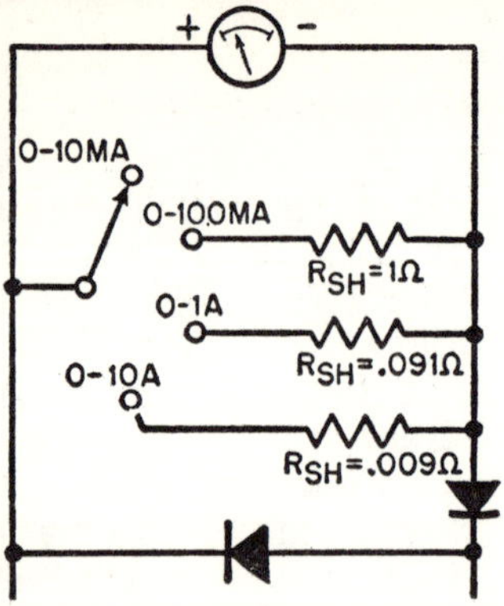

Fig. 1-4. The ac milliammeter.

difference, since the added 1 mA would have meant a deflection of 1/20 of the distance between the second and third subdivisions of the two major graduations. The added accuracy of several scales adds to the versatility of the meter, which allows the technician to use the same meter **accurately** in more applications.

The meter discussed up to this point works well for dc circuits but needs modification if alternating current is to be measured. In Fig. 1-4, a rectifier unit is used to convert the sine-wave input to a dc value which is representative of (though not equal to) the effective input. A meter scale is provided which is calibrated to read the effective, or rms, value of the sinusoidal current. Due to the expense of the added circuitry, many meters are not designed with the ac feature.

To measure current, the ammeter must be placed in the circuit in such a manner that circuit current flows through the meter as shown in Fig. 1-5. (When measuring dc, be careful to observe polarity when connecting the meter to the circuit.) Note that in each case the circuit is broken and the meter physically placed in the path of current—the negative probe to the side from which the current is flowing. Three different currents are present in the circuit: I_{R1}; I_{R2}; and I_t, the total circuit current.

The Voltmeter

The same basic meter movement used in the ammeter can also be used in the voltmeter. Large-value resistors, called **multipliers**, are placed in series with the meter to limit current flow. As in the ammeter, the values of these resistors are determined by the current needed for full-scale deflection of the pointer and by the voltages to be measured. Because the current through the meter will be proportional to the applied voltage, the scale can be calibrated directly in volts for a specific multiplier resistor.

As an example, let's use the 100-ohm, 100 uA meter movement and make a dc voltmeter with a full-scale reading of 1V. Since the coil resistance of the meter is 100 ohms and 0.0001A causes full-scale deflection, the total resistance of the meter coil and multiplier is

$$R_t = E / I = 1 / 0.0001 = 10{,}000 \text{ ohms}$$

The coil furnishes 100 ohms of the total resistance; therefore, the value of the series (multiplier) resistor is

$$10{,}000 - 100 = 9{,}900 \text{ ohms}$$

Figure 1-6 shows a dc voltmeter capable of reading voltages in four separate ranges. Multirange meters such as this use one meter movement and a group of multipliers connected through a switching arrangement.

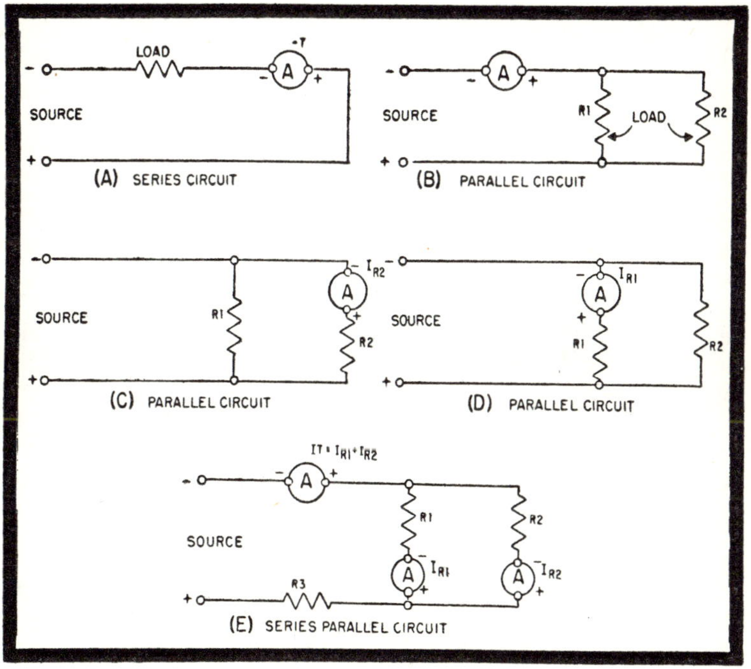

Fig. 1-5. Using the milliammeter.

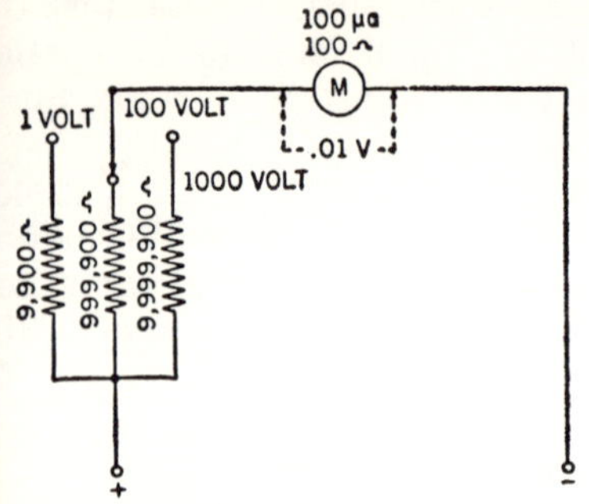

Fig. 1-6. The voltmeter.

Ac voltmeters, similar to ac ammeters, use a rectifier circuit to obtain a dc voltage proportional to the applied ac. The calibration of ac voltmeters may be in both rms and peak-to-peak (p-p) values.

The **sensitivity** of voltmeters is stated in **ohms per volt**, and is found by dividing the resistance of the meter coil plus the multiplier resistor by the full-scale reading in volts. For example, the sensitivity of the 100 uA, 100-ohm meter movement used with a 99,900-ohm resistor to read 10V full scale (Fig. 1-6) is

$$(99{,}000 + 100) / 10V = 100{,}000 / 10 = 10{,}000 \text{ ohms per volt.}$$

The sensitivity of a voltmeter determines the accuracy of the readings obtained. To read the voltage across a component, the meter is placed in parallel with the component. In order to read the voltage drop across R1 in Fig. 1-7, for example, the meter is placed in parallel with R1 as shown in B and C. While connected in this manner, a portion of the original current that flowed in R1 is shunted through the meter. The amount of current drawn by the meter is determined by the meter's sensitivity. A meter with a very high internal resistance (hence a high sensitivity) will have very little effect on the circuit. However, if the internal resistance of the meter is low or the circuit under test has a very high resistance, the meter will draw a large current, or "load down" the circuit, and the reading will not be accurate.

In Fig. 1-7, the circuit to be tested contains two 10K resistors with a total 10V applied; 5V is dropped across each resistor. When a meter with a sensitivity of 20,000 ohms per volt is placed across R2 and the range switch is set to **10V**, the

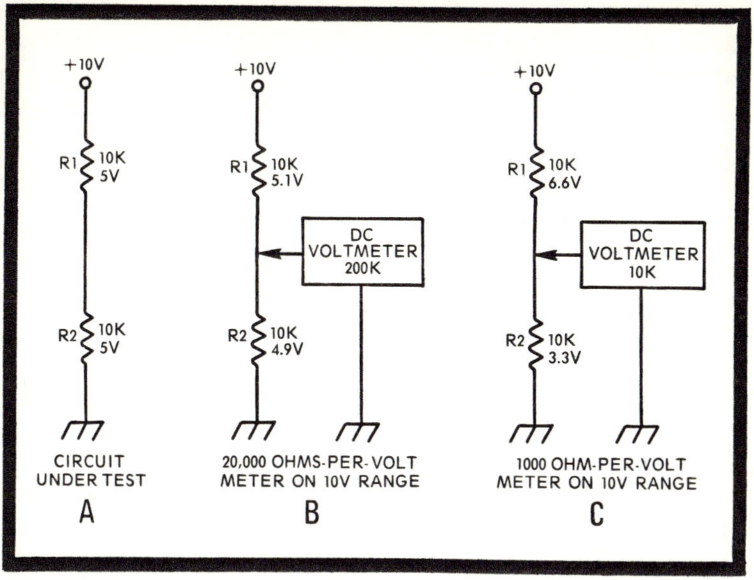

Fig. 1-7. Demonstrating the effects of meter sensitivity.

meter adds 200K in parallel with R2 (Fig. 1-7B). The power supply no longer "sees" a total resistance (R_t) of R1 plus R2, but a total resistance computed as R1 plus the combined resistance of R2 and the meter.

$$R_t = \frac{R1 + R2 \times R_m}{R2 + R_m}$$

The equivalent resistance of R2 and the meter in parallel is roughly equal to 9.5K. Therefore, the circuit will act as a 10K resistor (R1) in series with the 9.5K equivalent circuit. The total resistance will now be 19.5K instead of 20K. The current and voltage drop of R1 will increase, decreasing the voltage drop across R2.

If the meter sensitivity were 1000 ohms per volt, the circuit voltages would be as seen in Fig. 1-7C. This shows how the inaccuracies that occur due to circuit loading by the meter depend on meter sensitivity.

The Ohmmeter

Since resistance is the opposition to current, it can be measured by placing a known voltage across the resistance to be measured and monitoring the resulting current through the resistance. Since current equals voltage divided by resistance,

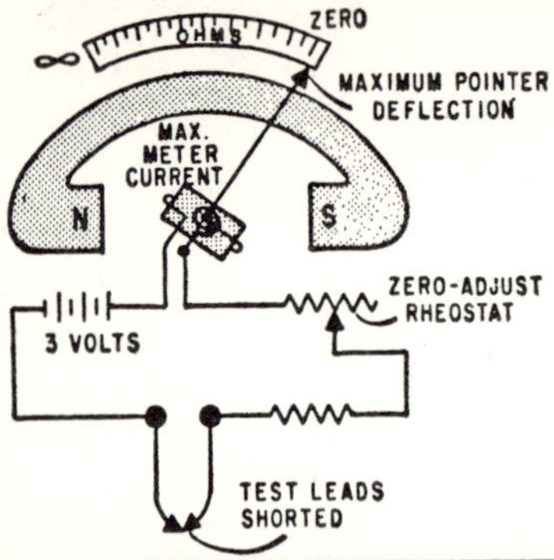

Fig. 1-8. The ohmmeter.

the greater the current, the less the unknown resistance must be. The typical ohmmeter consists of a milliammeter, a voltage source, and a resistor connected in series, as shown in Fig. 1-8.

Before the meter is used, the test leads should be shorted together and the variable resistor adjusted for full-scale deflection of the pointer, indicating a resistance of zero ohms (Fig. 1-8). When an unknown resistance is placed between the test leads, the meter reading decreases, showing a decrease in total current but an increase in resistance. The calibration of the scale is such that minimum resistance is indicated by maximum deflection of the pointer, and maximum resistance is indicated by minimum deflection. As with the other meters, a switching arrangement is used in practical ohmmeters to provide different resistance scales.

When using the ohmmeter, care should be taken to insure that no voltage is present in the circuit under test (other than that supplied by the ohmmeter) and that no parallel path for current exists. See what happens in Fig. 1-9 when there is more than one current path. Suppose you are to measure the resistance of R1. If the meter is inserted as shown in A, two paths are available—one through R1 and one through R2 and R3. In this case, if all three resistors were good, you would get a 5-ohm reading on the meter.

Figure 1-9B shows the proper method of reading the resistance of R1. Notice that one end of R1 is disconnected from the rest of the circuit. Now only one current path exits and a correct reading of 10 ohms can be obtained.

THE VOM

Having to have around three separate meters to make the basic electrical measurements probably led someone to invent the volt-ohm-milliammeter, or **VOM.** This combination instrument includes most of the features found in the three basic meters just discussed. Requiring no external power source, this piece of test gear is always ready to go to work in almost any location.

The VOM, or **multimeter**, as it is frequently called, is probably the most useful of all electronic test equipment. This is one of the first pieces of test gear a technician learns to operate. Units are available in most price ranges, depending on the sensitivity, accuracy, and features offered. The meters explained here illustrate some of the typical characteristics of VOMs.

Front-Panel Description

The photo in Fig. 1-10 shows some of the features found in a typical VOM. At the top of the panel a single meter movement with different scales for various functions is located under a transparent face. A single knob provides for selection of the desired use and adjustment of the meter circuit. Five test lead jacks permit the unit to be connected to a circuit for various measurements.

The **function** switch allows for the measurement of four different quantities: dc voltages (DCV), direct current (DCMA), resistance (Ω), and ac voltage (ACV). Notice that no provision is made to monitor alternating current. This is frequently the case, except in some of the more expensive meters.

The switch also controls the **range** of the meter. Twenty different ranges are provided. Two jacks at the lower left allow the range of the meter to be extended to 6000V.

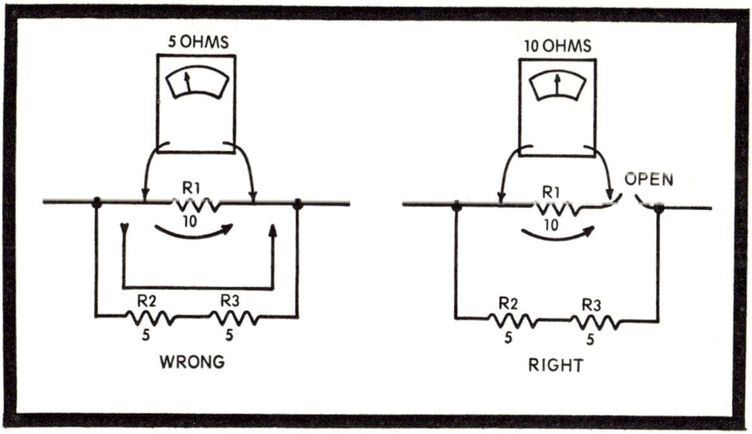

Fig. 1-9. Using the ohmmeter.

Fig. 1-10. A basic VOM. (Courtesy Triplett Corp.)

When the function switch is placed in any **DCV** or **ACV** position, the meter functions as a voltmeter. The range of the meter is established by the specific setting of the switch. Ranges of 3V, 12V, 60V, 300V, 1200V, and 6000V are available. Circuit voltages can be measured by connecting the common jack to one side of the circuit and the **V- Ω -A** jack to the other side. Voltages up to 6000V can be measured by using the **6000V dc** or **6000V ac** jack instead of the **V-Ω-A** jack. When this is done, the range switch should be in the **6000V dc** or **6000V ac** position.

Most ac and dc voltages are read from the next-to-the-top scale. With the switch in the **12V**, **60V**, **300V**, **1200V**, or **6000V** position, this scale is used for ac or dc voltages. Each subdivision between the numbers is equal to 2V on the **12V** scale, 10V on the **60V** scale, 50V on the **300V** scale, 200V on the **1200V** scale, and 1000V on the **6000V** scale. Small ac voltages are measured on a separate scale below the one described.

To read direct current, the meter is connected in the proper polarity by the **com** and **V- Ω -A** jacks. With the switch set to **DCMA** and the desired range, direct current in milliamperes can be read from the same scales used for voltages.

Resistance is measured using the Ω position of the **function** switch and reading the value of the resistance from the upper scale while the unknown resistance is connected between the **com** and **V- Ω -A** terminals. Notice that the resistance scale has its zero resistance mark at the full-scale deflection point, indicating maximum current. Four different resistance ranges are provided. In the **x1** position, the reading is taken directly from the scale. In all other positions, the reading is multiplied by the figure given on the index of the switch. For example, if the pointer comes to rest exactly on the "20" mark of the ohms scale with the switch in the **x1000** position, the actual resistance is

$$20 \times 1{,}000 = 20{,}000 \text{ ohms}$$

The graduation of the resistance scale is quite different from that of the voltage and current scales. While the latter scales are marked with almost equal divisions, the marking of the ohms scale is not linear. This can be explained by reference to Fig. 1-11. Assume that you have an ohmmeter

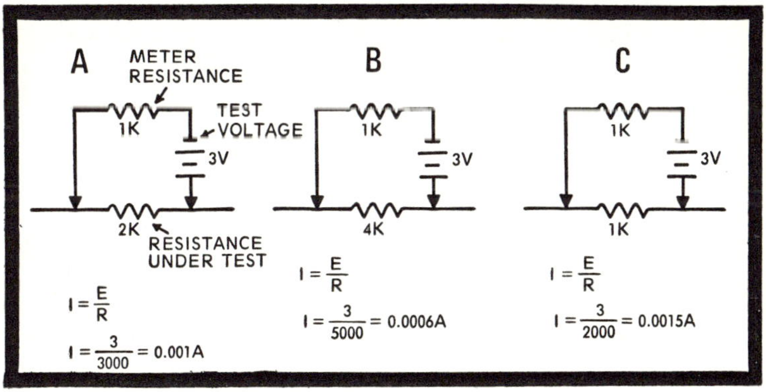

Fig. 1-11. The effects of resistance on meter current.

with a built-in resistance of 1000 ohms on some particular scale and a test voltage of 3V. These values remain constant as long as the scale switch remains in one position.

In Fig. 1-11A, a 2K resistor is connected across the meter. Current through this series loop is calculated as voltage divided by resistance, or 0.001A. Notice in B that when the resistance under test has been doubled the circuit current does not decrease by half as might be expected. Although the test resistance was doubled, the total resistance (test resistance plus meter resistance) was not doubled. Neither will decreasing the test resistance by one-half decrease the total resistance by a similar amount. Hence, a linear change in the resistance being tested does not cause a corresponding linear change in total current; therefore, the resistance scale cannot be linear.

Meter Operation

The operation of a meter was discussed when we looked at the front panel of the VOM; however, there are a few more important features that should be mentioned. These also will be pointed out when we look at the electronic voltmeter.

Before using the VOM to make any resistance check, you should first check the meter zeroing on the scale to be used. A scale should be selected that will yield a reading at approximately center scale, for greatest accuracy. For example, assume that you want to measure the resistance of a device of about 75 ohms, using the meter in Fig. 1-10. The **x1** scale could be used, but the pointer would be deflected so far to the left that the reading would be somewhat inaccurate. Also, each small graduation between "50" and "200" represents 50 ohms—not very precise. By shifting to the **x10** scale, 75 ohms is represented by the "7.5" mark, between "5" and "10," and a more accurate reading is possible.

Zeroing is accomplished by selecting the correct scale and then touching the probes together. Use the **ohms adj** control to set the pointer at zero resistance. If the meter will not zero, try replacing the batteries. When the circuit resistance is measured and the reading falls way above center scale, set the meter to a higher range. If the reading falls far below center scale, select a lower range.

An ohmmeter should never be connected to an energized circuit. The voltages could result in damage to the meter. Also watch out for charged capacitors—especially those large electrolytics...they should be discharged before the meter is connected to the circuit. Again, insure that there are no other resistances in parallel to upset the reading.

Always deenergize a circuit before attempting to measure its current. After the power has been removed, connect the meter in series in the circuit. Make sure that the ammeter function has been selected and the meter is set to the proper scale. If dc is being measured, you'll have to watch the lead polarities too—and **never** measure ac with the dc function selected, or vice versa. When all is well, energize the equipment to read the current.

Voltage measurement involves the same precautions as current measurement. Don't try to measure dc voltage on the ac voltage function, and use the correct scale. When measuring voltage, start with a scale well above that of circuit voltage. Then when you measure the voltage, the meter will not be damaged by excessive voltage. To achieve the desired accuracy, the meter can be set back, one range at a time, until a range is reached where a center-scale reading is obtained.

Caution: Never connect a voltmeter to an energized high-voltage circuit. First deenergize the circuit, discharge the capacitor, and establish the proper connection: **then** energize the equipment. Holding a meter while making high-voltage measurements can result in a fatal shock if the meter shorts out or an arcover occurs.

Now that you have a general idea of how to operate a VOM, let's look at some more of the features that are available in various units.

Comparing Multimeters

By comparing the specifications given for two meters (Table 1-1) we can learn a little more about some of the

Table 1-1. Comparison of VOMs.

	Meter A	Meter B
Sensitivity	1K (ac/dc)	(30K dc/15K ac)
AC Volts	0 - 15	0 - 2.5
	0 - 150	0 - 10
	0 - 1K	0 - 25
		0 - 100
		0 - 250
		0 - 500
		0 - 1K
DC Volts	0 - 15	0 - 0.5
	0 - 150	0 - 2.5
	0 - 1K	0 - 5
		0 - 25
		0 - 50
		0 - 100
		0 - 250
		0 - 500
		0 - 1K
DC Milliamperes	0 - 150 mA (one scale)	0 - 0.05 mA
		0 - 5
		0 - 50
		0 - 500
Resistance	One scale: 0-100K, 2.5K center scale	Rx1: 0-2K, 30-ohm center scale
		Rx100: 0-100K, 3K center scale
		Rx10K: 0-1M, 300K center scale
AC Amperes	None	None

features of the VOM. Notice in the specifications that the sensitivity of meter B is 30 times that of meter A on dc scales and 15 times that of A on the ac scales. The greater sensitivity of B causes much less circuit loading and, therefore, a much greater accuracy.

Although both meters have ranges to 1000V, only three scales are provided for meter A. The greater number of scales on B allow for scale selection more compatible with the voltage being measured. This too will improve accuracy. The low dc voltage ranges allow for accurate transistor bias voltage readings.

Not only are more dc milliampere ranges provided on meter B, but also a much wider total range. A similar situation exists for the resistance sections. By the way, I have both of these meters, and the lowest resistance I can read with any accuracy on meter A is about 50 ohms. Note that neither meter has any provision for reading alternating current.

What else is different about these meters? Price! Meter A cost me $6, B about $40. So why waste $6 on A? Although it doesn't have the versatility or accuracy of the other meter, it still offers a few features I like. It's about the size of a package of cigarettes and doesn't weigh much more. A lot of my work involves automatic controls for electrical machinery. This lightweight little instrument fits in nicely with wrenches, screwdrivers, and pliers in a small troubleshooting kit I carry for these jobs—and it's still accurate enough for most purposes.

THE EVM

The meters discussed up to this point are all nonelectronic—that is, they use only magnetism and electromagnetism to measure the basic quantities of voltage, current, and resistance. Herein lies the difference between the VOM and electronic voltmeters (EVMs). The EVM is similar to the VOM in appearance, but differs considerably inside. Most EVMs have provisions for resistance and voltage measurements, and some of the better units provide for current and stage gain measurements.

The same features can be obtained in a VOM, but the added electronic circuitry within the EVM adds to the accuracy of the meter. Early models of the EVM were vacuum-tube devices (VTVMs), and there are still a lot of these "old guys" around. Now, with extensive use of solid-state circuits in test equipment, many high-quality instruments contain no vacuum tubes; hence we speak of EVMs instead of VTVMs. The term "EVM," as used here, includes all portable, electronically operated test equipment designed to measure current, voltage, or resistance...this means the VTVM, FETVM (field-effect transistor voltmeter), the TVM (transistor voltmeter), and any other meter that uses amplifiers to isolate the basic meter from the circuit under test.

THE VTVM

By now it should be evident that when a meter is connected into a circuit, the meter will draw current from the circuit under test, and the amount of current drawn will depend on the meter sensitivity. This cannot be avoided and, in many cases, is a problem. In fact, some circuits are so critical that the slight current drawn by the meter will upset circuit operation to a point that a usable indication cannot be obtained. For this reason, the VTVM is often used to measure voltages in instances where circuit loading cannot be tolerated.

DC Voltage Measurements

To see what reduces the loading effects of the VTVM, look at Fig. 1-12. Such a circuit can be used to measure the dc voltage present in a circuit and keep circuit loading to a minimum.

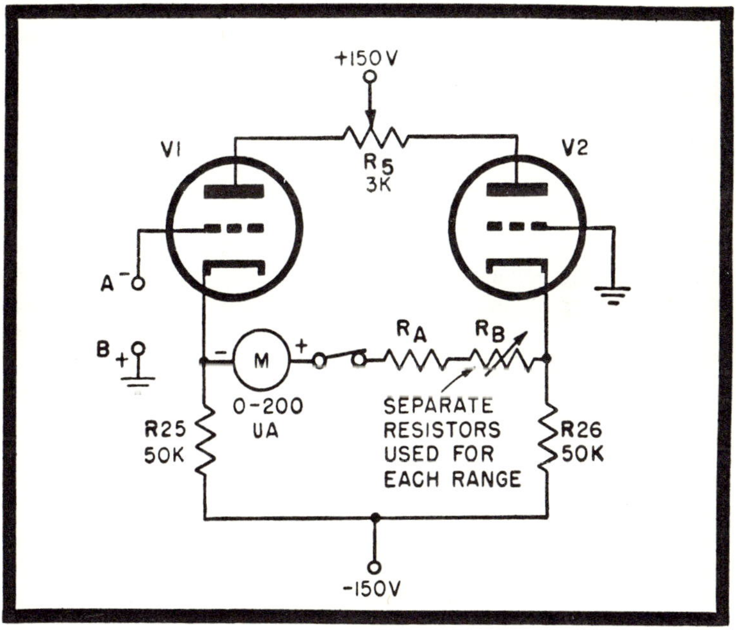

Fig. 1-12. The basic dc VTVM circuit.

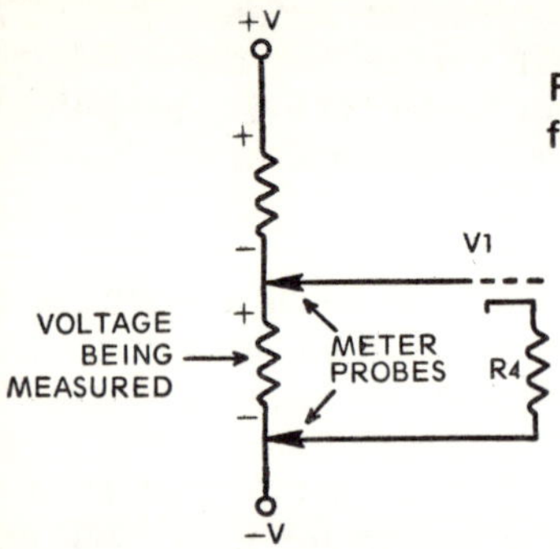

Fig. 1-13. Circuit-loading effects of the dc VTVM.

The meter probes connect the grid and ground points of V1 across the voltage to be measured. The input impedance of a vacuum tube is very high, usually several megohms for those used in VTVMs. Figure 1-13 represents the parallel path of current established when the meter is connected to the circuit under test. Since current would have to flow between the cathode and grid elements of V1, the input sensitivity is kept high as long as the input voltage does not cause appreciable grid current to flow. Referring back to Fig. 1-12, you see that the meter movement itself is located in the plate circuit of V1; therefore, only changes in plate voltage affect the meter reading. This results in isolating the circuit under test from the meter proper.

However, since the plate voltage of V1 will respond to changes in voltage at the grid, the meter can still be made to indicate circuit voltage. In this simplified example (Fig. 1-12), the meter leads are held open and R1 adjusted until the meter reads zero volts. This adjustment balances the circuit to eliminate any errors caused by the inherent differences of the conduction of V1 and V2. Connected in the manner shown, the meter actually measures the difference between the plate voltages of V1 and V2.

Assume that the circuit shown is receiving plate voltage and no connection exists between the meter leads, points A and B. The tubes conduct across their respective load resistors and halves R5 to the B+ supply at the wiper. If identical vacuum tubes are used, then both halves of the circuit will conduct equally and the plate voltages will be the same. Small inconsistencies in the manufacture of the components will cause a slight different in conduction; therefore, R1 is used to raise the plate supply voltage of one triode while lowering that of the other, to achieve a balanced condition at the plates. With a balanced condition at the plates, there is no potential difference and no current flow through the meter.

When a voltage is supplied between points A and B, this balance is upset. The plate voltage at V1 will either increase or decrease, depending on the polarity of the applied voltage. When the plate voltage at V1 changes, there is no change at V2, and a potential difference exists, causing current to flow through the meter.

AC Voltage Measurements

The troubleshooter is primarily interested in two types of ac voltage measurements—peak-to-peak and effective (rms). Effective voltage readings can be made with a rectifier built into the input of the dc VTVM circuit shown in Fig. 1-12. Effective voltages are of interest in many troubleshooting applications. However, for circuits containing complex waveforms, it is sometimes much more desirable to know the maximum positive and negative excursions of the waveform, or the peak-to-peak voltage. A circuit capable of such measurements is shown in Fig. 1-14.

Two diodes are connected in series with the meter in voltage-doubler fashion. During the first half-cycle of the applied voltage, point A is positive with respect to point B. The positive voltage at the right plate of C1 will attract electrons from the plate of V2, while the negative voltage at the cathode of V2 causes V2 to conduct. Under this condition, C1 is furnished charging current by electrons leaving point B and traveling through V2 to one plate of C1. Electrons leave the opposite C1 plate via point A to the circuit under test, travel through the circuit, and return to point B. The result is that during the first half-cycle of operation C1 charges in a manner that yields a negative voltage at the cathode-plate junction of V1 and V2.

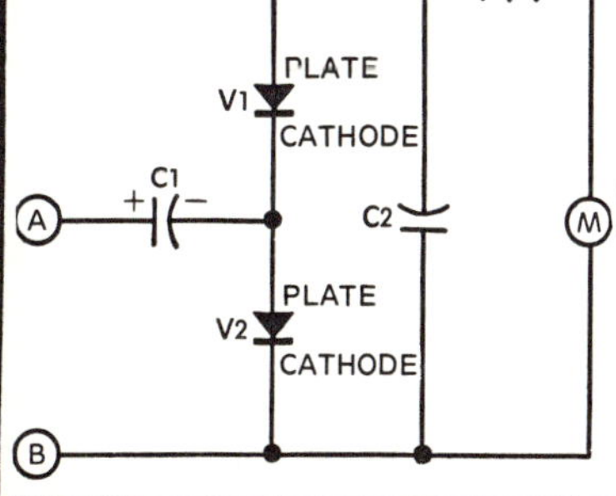

Fig. 1-14. A peak-to-peak VTVM.

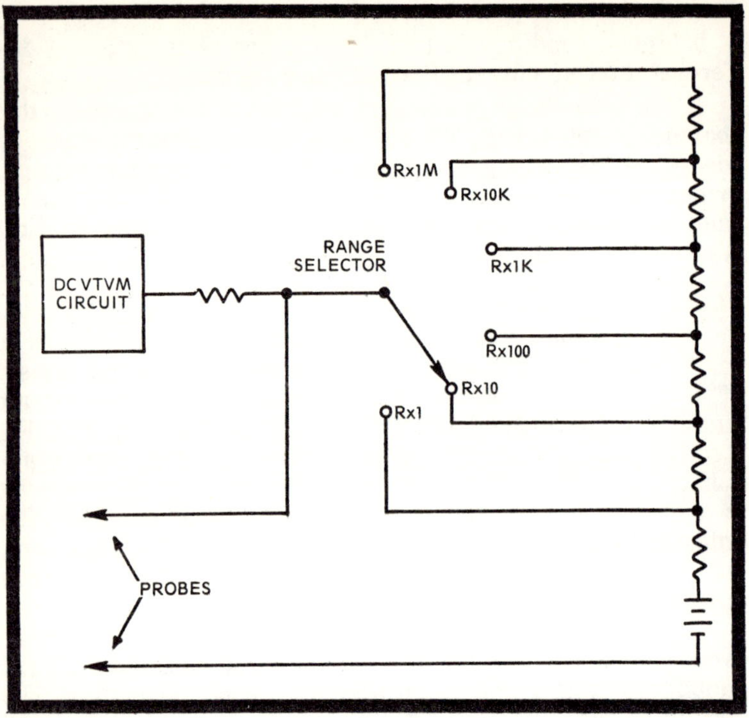

Fig. 1-15. A VTVM with ohmmeter capabilities.

During the second half-cycle (point A negative), V1 "sees" the charge across C1 in series with the applied circuit voltage at point A. Tube V2 has a negative potential at the plate and, therefore, will not conduct. Tube V1 conducts, its current flowing through the resistor and the meter movement to point B. Capacitor C2 charges to the voltage drop across the resistor and meter movement. Under ideal conditions, this charge would be equal to twice the charge across C1. Capacitor C2 is charged only on alternate half-cycles of the applied ac, and C2 is always attempting to discharge through the resistance and the meter.

Resistance Measurements

Like the VOM, the VTVM can also be designed to measure resistance; such a circuit is shown in Fig. 1-15. According to the setting of the **range selector** switch, any one of several resistors connected in series with the test voltage may be connected to the circuit under test. This arrangement permits the full battery voltage to act as the bias voltage for the dc

VTVM circuit when an infinite resistance (probes open) is between the probes. With the probes open, the meter is adjusted so the pointer is at full-scale deflection. As resistance between the probes is decreased, the bias voltage to the meter circuit is decreased, in turn decreasing point deflection. The upper range of such a meter is often an Rx1,000,000 range. The VTVM-type ohmmeter can measure a higher resistance with a lower test voltage than the VOM. The VOM resistance-measuring capability is limited by the sensitivity of the meter movement and the internal test voltage, while that of the VTVM is limited by the gain of the vacuum tube and the sensitivity of the meter.

Another advantage of this type of circuit is that there is no need for initial adjustment after warmup, regardless of how often the ranges are switched. The adjustment is made in the dc VTVM circuit, and is therefore not a part of the resistance function.

The Solid-State EVM

Electronic voltmeters have not escaped the solid-state revolution. In fact, the compact, rugged, lightweight transistor has been found to be the ideal component for test in-

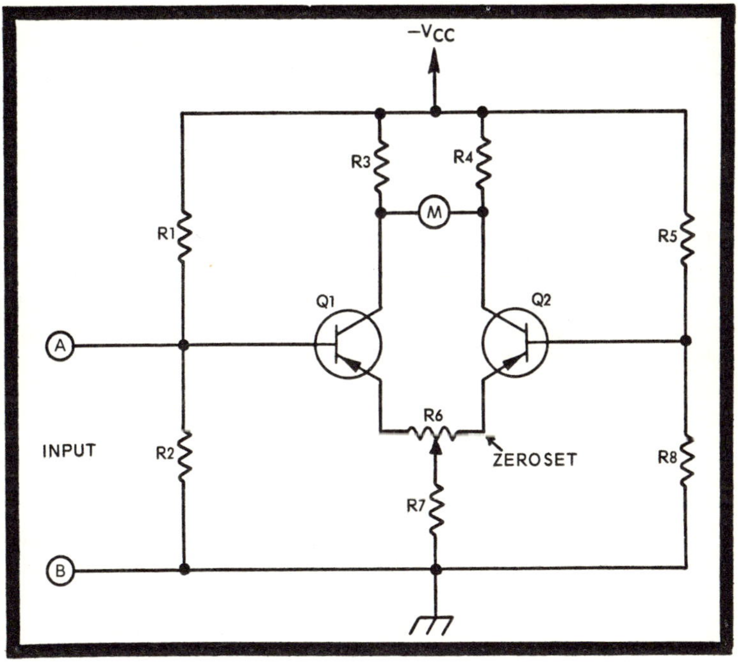

Fig. 1-16. A transistor dc EVM.

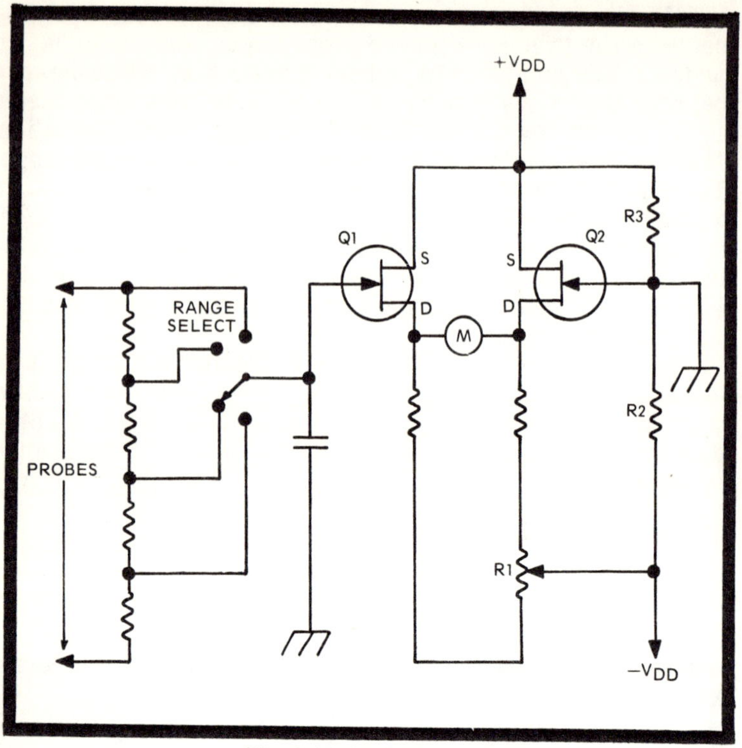

Fig. 1-17. An FET dc EVM.

struments. Some of the meters with solid-state components have features not seen in the VTVM.

The basic dc VTVM circuit shown in Fig. 1-12 has its counterpart in the solid-state version shown in Fig. 1-16. This, like the vacuum-tube model, uses a **differential amplifier**. This amplifier uses two transistors connected to a common stabilizing resistor, the current flow through both transistors developing a common bias. Under static conditions with no input to point A, the outputs at the two collectors are equal and no current flows through the meter. When an input is present at point A, the circuit becomes unbalanced and more current flows in one transistor than the other, causing unequal collector voltages and a current through the meter. In practice, the emitter resistor R7 may be replaced by a constant current source in the form of a transistor-regulated supply.

Actually, the FET circuit shown in Fig. 1-17 is the preferred version. By using the FET circuit, it is possible to obtain input impedances in excess of 100M. Resistor R1 in the drain circuit allows for zero setting the meter. A constant gate bias is supplied to Q2 by voltage divider R2R3. With the ex-

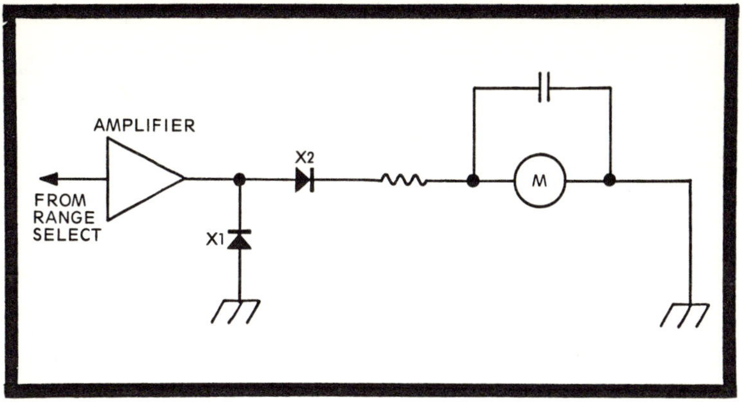

Fig. 1-18. A solid-state ac EVM.

ception of the FETs, this circuit operates similarly to meter circuits previously discussed.

Figure 1-18 illustrates how the solid-state EVM may be adapted to measure ac voltage. The voltage to be measured is fed through the solid-state amplifier circuit and coupled by C2 to the diode circuit, X1 and X2.

From this brief description we can see that the major difference between the solid-state EVM and the VTVM is in the nature of the amplifiers and rectifiers themselves; basic operation is the same for both units.

EVM Features and Accessories

We have already discussed the input impedance advantage of the EVM over the common VOM; however, there are other important advantages. Increased frequency response is one. Most VOMs are accurate to maybe a few hundred hertz. Above this, frequency response falls off and accuracy is questionable. By using a special probe, an EVM can measure ac voltages well into the megahertz range. With components of proper values, the test lead shown in Fig. 1-19 can extend the operating range of one popular make of EVM to 250 MHz. This lead is connected to the dc input of the meter

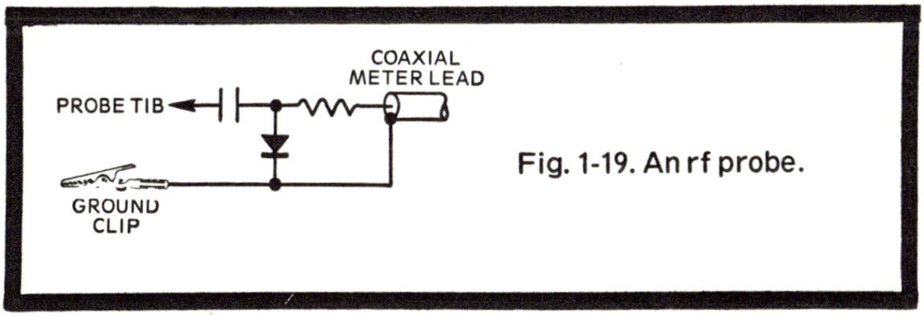

Fig. 1-19. An rf probe.

and the meter set up for dc operation. The rms value of the rf voltage can then be read directly from the dc scales.

The dc high-voltage range of the EVM can also be extended by use of special probes. These probes are special safety probes with built-in high resistances, allowing for dc voltage measurements up to 30 kV.

Due to the EVM's inherent ability to "amplify" very low voltages, the upper resistance range is extended. Another added feature of EVMs is the provision for low-power resistance readings. This characteristic allows for in-circuit resistance measurements by using a very low test voltage—more about this later.

Other scales usually added to the EVM include decibel and zero-center scales. These too will be discussed in greater detail later.

EVM Operation

2

The operation of the EVM can be best illustrated using an actual meter as an example. The meter used here for such demonstrations is the FE20 HI-LO field-effect meter manufactured by Sencore (Fig. 2-1). In my view, this is one of the top EVMs on the market for general service use. Priced under $150, this meter offers the versatility and accuracy necessary in a portable service instrument.

EVM DESCRIPTION

The panel of the FE20 is arranged for ease of operation. Two switches provide for a quick test of the batteries, with a scale on the meter panel provided to show proper voltage levels. Two separate power supplies are used, the main supply (9V) and the ohmmeter supply (1.5V). Overcurrent protection is furnished by a 0.6A fuse on the front panel. The **zero adj-zero ctr adj** and the **ohms adj** controls allow for initial zeroing of the meter movement.

The meter movement has six scales in addition to the battery test scale. A mirror is placed below the second scale to aid in obtaining accurate readings by eliminating any **parallax** (viewing angle) error. A separate scale is provided for each of the following functions: ohms, dc volts (combined with dc milliampere scale), dc volts center scale, ac volts rms, ac volts peak-to-peak, and decibels. Two switches control the mode of operation and scale selection.

A 6 in. 200 uA d'Arsonval meter is used for the movement. This meter is accurate within 2 percent and is protected by a shunt diode. The circuitry is completely solid state, with most of the electronics contained on one printed-circuit board.

The **function** switch is a 7-position switch that selects one of seven modes of operation: —dc volts, +dc volts, ac volts, high-power ohms, low-power ohms, +dc mA, or —dc mA. The **range** switch has 12 positions controlling 12 dc voltage scales, 12 dc voltage scales referenced to center scale, 9 rms scales, 9 peak-to-peak scales, 7 resistance scales, and 9 dc milliampere scales.

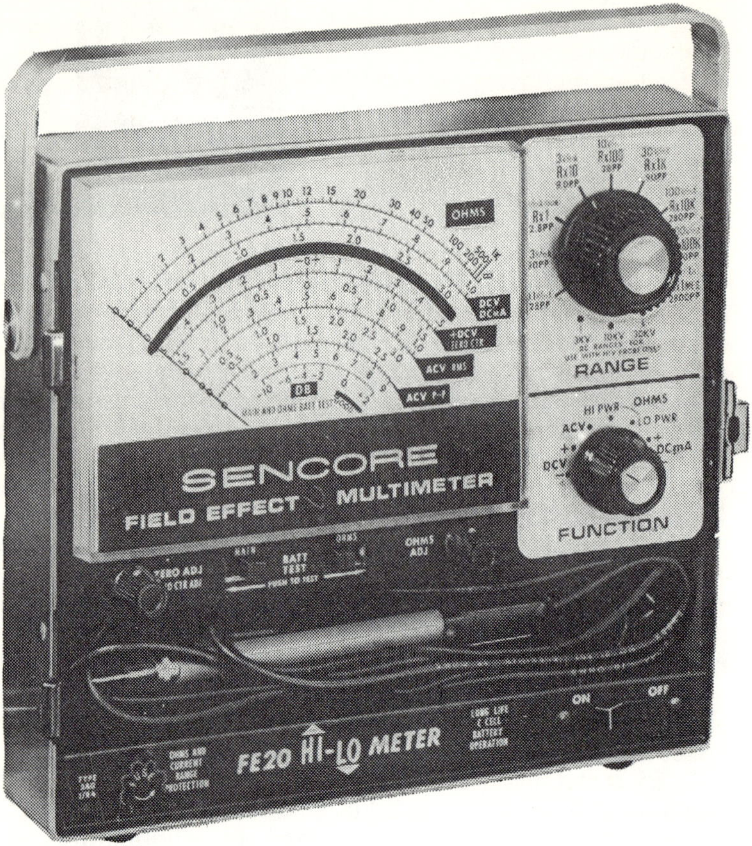

Fig. 2-1. The FE20 field-effect meter. (Courtesy Sencore.)

The input resistance in the dc mode is 15M. A 120K switchable isolation resistor is built into the probe. Accuracy is 1.5 percent with 30 dB minimum ac rejections. The ac input impedance is 12M shunted by 90 pF with a ±1 dB frequency response between 10 Hz and 150 kHz, dropping to ±3 dB between 5 Hz and 500 kHz. Accuracy in the ac mode is 3 percent.

The entire unit including test probes is contained in a compact, vinyl-clad, steel carrying case. A high-voltage test probe that extends the upper dc range to 30 kV is contained in the lid of the case, while the normal test leads are attached to the body.

A schematic of the Sencore FE20 is shown in Fig. 2-2. A balanced-bridge circuit consisting of two FETs and two bipolar transistors provides a high input impedance and excellent stability. When measuring ac voltages, an additional amplifier stage, TR5, is used to drive the peak-to-peak detector, CR3 and CR4.

The dc voltage at the input passes through the ac rejection filter, consisting of R1 and C1, to the voltage divider, R2, through R10. **Range** switch S3B selects the required input voltage to be applied to TR1. Transistor TR1, being an FET, offers a very high input impedance and drives the base of TR2. Both collector current from TR2 and source current from TR1 flow through R29; therefore, the outputs of these two transistors are in parallel. Conditions in the TR3 and TR4 section of the balanced bridge are held constant during this mode of operation, because the gate of TR4 is tied to ground through the **ac calibration** control, R36. The meter, in series with the **dc calibration** control, is connected between the collector of TR2 and the collector of TR3, the positive meter lead being connected to the collector of TR2. Any change in temperature will equally affect both sides of the bridge circuit. This results in a very stable circuit.

When measuring direct current, one of the standard resistors, R17 through R25, is selected by the **range** switch and placed in series with the circuit current. The dc voltage-measuring circuits are then used to measure the voltage across the standard resistor. Fuse, F1 protects the resistors up to the 300 mA range. In addition, the meter movement itself is protected on all ranges by the isolation of the bridge circuit and the shunt furnished by CR5.

In the **hi pwr** resistance mode, the value of an unknown resistor is measured by connecting the resistor, a 1.5V battery, and a standard resistor in series. The voltage across the unknown resistor then is measured. The **ohms adjust** has the same effect as the **dc calibration** control, in that it is connected in series with the meter and is used to adjust meter sensitivity. When there is an infinite unknown resistance (open circuit), the FE20 will measure full battery voltage. The **ohms adjust** control is adjusted with the test leads open to read **infinity** on the meter. When the leads of the FE20 are shorted together, the voltage will be zero, and the meter will read **zero**. If the unknown resistance is exactly equal to the standard resistance, the voltage across the unknown resistor will be one-half the total, and the meter will read one-half of full scale. The A-section of S3 (**range switch**) is used to select different standard resistors for different ranges.

The **lo power ohms** function of the FE20 is used when working on solid-state equipment where resistance can be

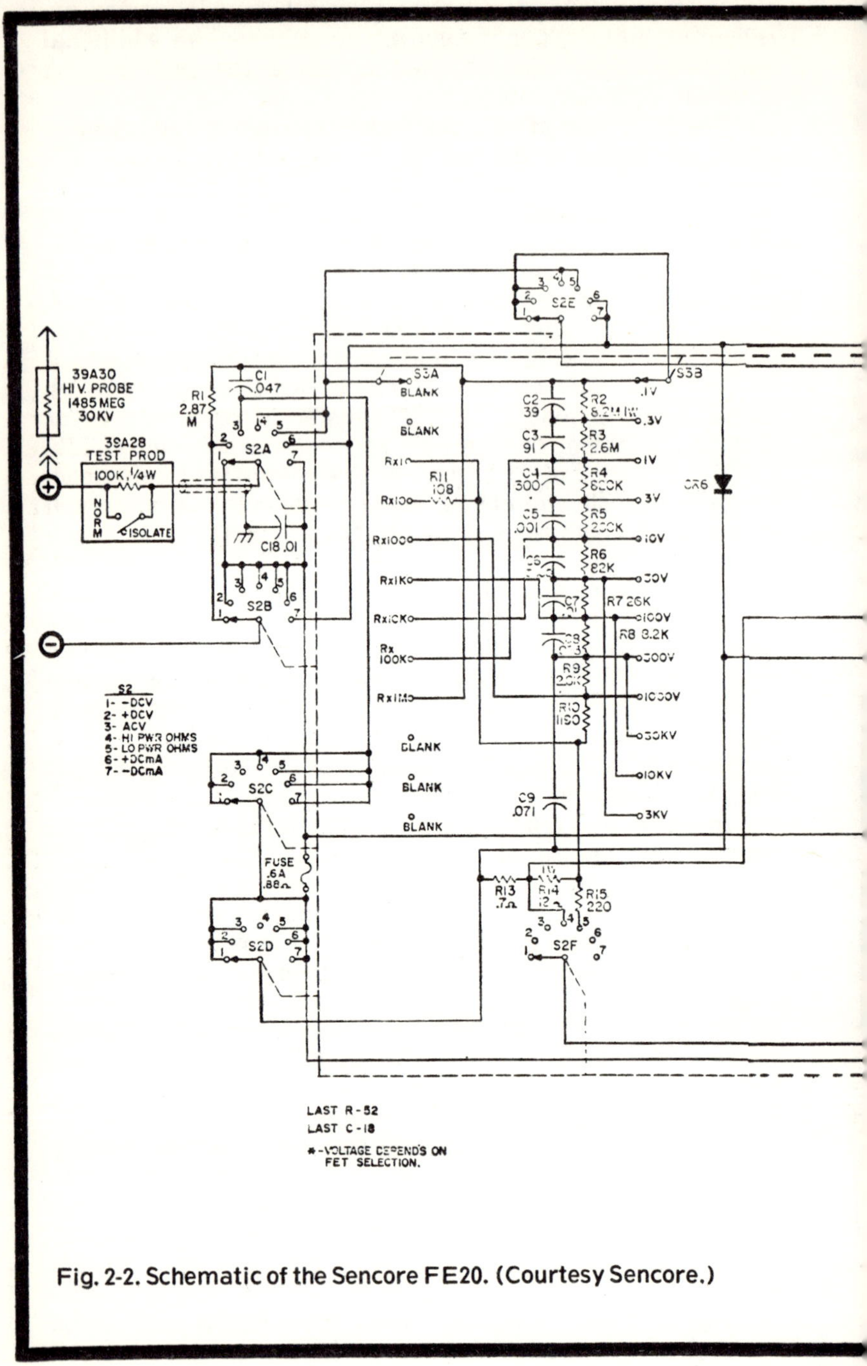

Fig. 2-2. Schematic of the Sencore FE20. (Courtesy Sencore.)

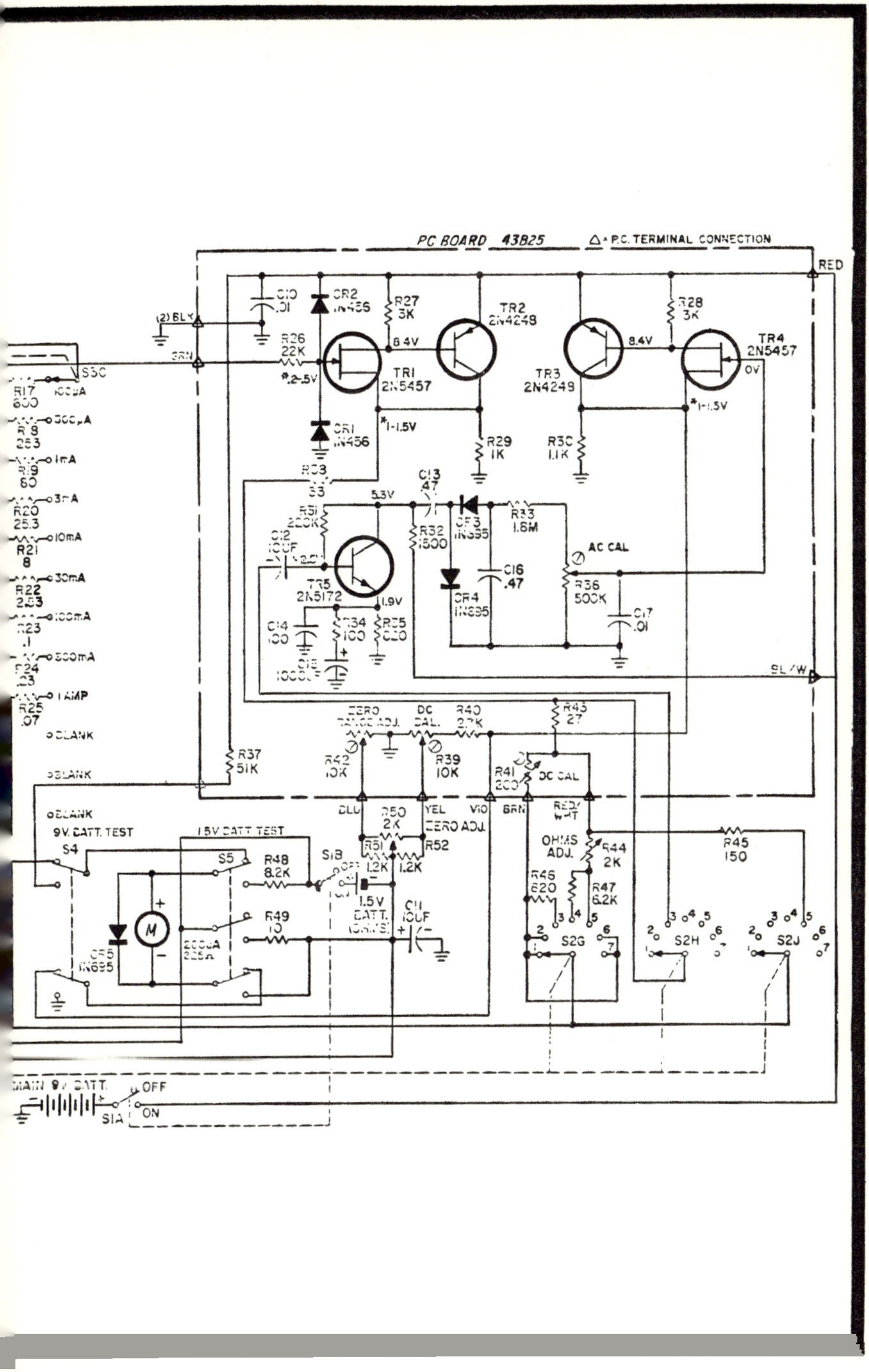
PC BOARD 43B25
△= P.C. TERMINAL CONNECTION
RED
C10 .01
CR2 1N456
R27 3K
TR2 2N4248
R28 3K
(2) BLK
GRN
R26 22K
*.2-.5V
8.4V
TR1 2N5457
TR3 2N4248
8.4V
TR4 2N5457
0V
*1-1.5V
*1-1.5V
S3C
R17 600
100μA
R18 253
300μA
1mA
R19 80
3mA
R20 25.3
10mA
R21 8
30mA
R22 2.53
100mA
R23 .1
300mA
R24 .23
1 AMP
R25 .07
CR1 1N456
R29 1K
R30 1.1K
R33
C13 .47
5.3V
R51 220K
C12 10UF
2.6V
TR5 2N5172
1.9V
R32 1500
CR3 1N695
R33 1.8M
AC CAL
C16 .47
CR4 1N695
R36 500K
C17 .01
C14 100
R34 100
R35 820
C15 1000UF
BL/W
ZERO RANGE ADJ.
DC CAL.
R40 2.7K
R43 27
R37 51K
R42 10K
R39 10K
R41 200
DC CAL
BLANK
BLANK
BLANK
9V. BATT. TEST
1.5V BATT TEST
S4
S5
R48 8.2K
S1B
OFF
BLU
R50 2K
YEL
ZERO ADJ.
VIO
BRN
RED/WHT
R51 1.2K
R52 1.2K
1.5V BATT. (OHMS)
OHMS ADJ.
R44 2K
R45 150
R46 820
R47 6.2K
R49 10
C11 10UF
CR5 1N695
M
200μA
S2G
S2H
S2J
MAIN 9V BATT.
S1A
OFF
ON

measured without disconnecting one end of the resistor. In the **lo power** mode, the FE20 applies only 0.08V (80 mV) to the circuit so that the diode junctions of a transistor will not conduct. This allows for many in-circuit tests not possible with a conventional VTVM.

This mode operates the same as the **hi power ohms** function, except a voltage divider comprised of R14 and R15 is used to set the test voltage to 0.08V. The dc voltage-measuring circuit is still connected across the unknown resistance, but the sensitivity of the voltage-measuring circuit is increased so that full-scale deflection now represents 0.08V.

An ac voltage is measured by coupling the input through C1 to the top of the voltage divider comprised of R2 through R10. The B-section of S3 selects the voltage to be applied to the input of TR1; that transistor acts as a source follower. The output of TR1 and TR2 is coupled through C12 and R38 to the input of TR5. Transistor TR5, connected as a common-emitter amplifier, provides a voltage gain of about 10, driving the peak-to-peak detector, CR3 and CR4. A negative output from the peak-to-peak detector is applied through R33 and the ac calibration control to the gate of TR4, causing an unbalanced condition in the bridge and a deflection of the meter pointer.

GENERAL OPERATION

The operation of the FE20 is very simple; merely select the desired test with the **function** switch and the appropriate range with the **range** switch, connect the test leads to the points to be measured, and observe the indication on the meter.

The meter is equipped with a shielded test lead to allow accurate measurements of ac in the presence of strong fields. This special probe has a 120K isolation resistor to isolate the cable capacitance when making dc voltage measurements in critical circuits such as rf, i-f, and sweep circuits. For all normal readings, including dc voltage, the probe switch is left in the **normal** position. The only time that the switch is placed in the **isolate** position is when it is necessary to isolate cable capacitance from the circuit being measured. When the switch is in the **isolate** position, the reading will be only 0.6 percent low. The **isolate** position should only be used with dc and ac voltages at 60 Hz—much greater error will result on other functions.

Measuring DC Voltage

To measure dc voltages, set the probe switch to the **normal** position, the **function** switch to either **+dc** or **—dc**, and the

range switch for a meter indication as close to full scale as possible. When measuring unknown voltages, always start with a range you know is greater than the circuit voltage. Once an indication has been achieved, lower the **range** switch setting until the proper scale is reached.

On the FE20, dc voltage is read on one of the two black scales just above the mirror on the meter. The upper scale is calibrated from zero to 1.0V, with each small scale division representing 0.02V. This scale is used when the **range** switch is set to the 0.1V, 1V, 10V, 100V, or 1000V scale. The lower dc scale is calibrated from zero to 3.1V, each small scale division representing 0.1V. It is used with the **range** switch set to the 0.3V, 3V, 30V, or 300V range. The 1.0V and 3.0V scales are read directly when the **range** scale is set to the **1.0V** or **3.0V** position. When reading the 0.1V, 10V, 100V, or 1000V scale, the reading is multiplied by the range.

For example, assume that the **range** switch is set for 100V and the reading is 0.83. The actual voltage is 83V. A similar example can be used with the other dc scales. If the **range** switch is set to the 0.3V, 30V, or 300V scale, the reading is multiplied by 0.1, 10, or 100. For a reading of 2.7 with the **range** switch in the **0.3V** position, the circuit voltage would be 0.27V.

Measuring DC Volts With a Zero-Center Scale

The zero-center scales are used to read plus or minus voltages without the necessity of changing polarity with the leads or the **function** switch. This mode of operation is especially handy when adjusting the output of a circuit for zero volts dc. The FE20 can be set up for zero-center-scale operation, connected to the output, and the circuit adjusted to read zero volts at the center scale.

To use the zero-center scales, set the probe switch to the **normal** position, the **function** switch to **+dc volts**, and adjust the **zero adj** control until the meter pointer is positioned over the zero mark at the center of the **DCV center** scale (just below the mirror). This gives ranges of ±0.5V (upper scale) or ±1.5V (lower scale). The meter scales are then multiplied by 0.1, 10, 100, or 1000—depending on the setting of the **range** switch. If the **range** switch is on the **300V** position, then the 1.5V scale is multiplied by 100, resulting in a range of ±150V. If the **range** switch is on the **0.1V** position, the plus or minus range is read as ±0.05V. Values to the right of the center are positive, while negative values are to the left.

Measuring DC High Voltages

High voltage is always dangerous, and caution should be observed when measuring high voltages. Be sure that the ground lead from the meter is connected to the appropriate ground in the equipment being measured. If the ground lead is not connected before the test probe is brought in contact with the high voltage, you are merely extending the high voltage closer to you. If the ground lead is connected to the wrong terminal, you run the risk of possible damage to the equipment being tested and present a danger to yourself. Keep clear of all metal surfaces that could bring you in contact with a high voltage or cause you to be grounded. It is always a good idea to deenergize a high-voltage circuit before connecting the meter to the circuit. Discharge all capacitors with a high-voltage shorting probe and connect the ground lead of the meter to the proper terminal. Now connect the proper test probe of the meter to the high-voltage point to be measured; then energize the equipment. After the voltage has been measured, again deenergize and discharge the circuit before removing the meter leads.

The 39A30 high-voltage probe supplied with the FE20 extends its range to 30,000V dc. This probe may be used with any dc voltage range of the FE20 to act as a x100 multiplier and to increase the input impedance of the meter to 1500M. The 0.1V range becomes a 10V range with an impedance high enough not to affect the most critical circuit.

To use the high-voltage probe, remove the two parts from the retaining brackets in the cover. Insert the probe banana plug in the banana jack inside the handle. Plug the test probe of the FE20 into the small opening of the high-voltage probe and place the probe switch in the **normal** position. To read the actual voltage, the **function** switch is set to the **3 kV, 10 kV**, or **30 kV** position—depending on the voltage to be measured. Multiply the readings on the 3V scale by 1,000 for the 3 kV range, on the 1V scale by 10,000 for the 10 kV range, and on the 3V scale by 10,000 for the 30 kV range. Never use the 3 kV, 10 kV, and 30 kV modes except with the 39A30 probe attached.

Measuring Direct Current

When measuring direct current with the FE20, the probe switch is set to the **normal** position, the **function** switch to the **+dc mA** position, and the **range** switch to the desired current range. Remove power from the equipment to be tested and connect the meter in series with the current to be measured. Make sure that the negative meter lead is connected to the negative source, and the positive lead is attached to the positive source of current. Reconnect the power to the equipment and read the correct current on the appropriate meter scale. Should the meter read in the reverse direction, the current polarity is reversed from that expected and the **function** switch should be turned to the **—dcmA** position.

The zero-center scales may also be used to measure direct current. With the **function** switch set to **+dcmA** and the range switch set correctly, adjust the meter for zero center and connect as discussed earlier.

Measuring AC Voltage

For ac voltage measurements, the **function** switch is set to the **ac volts** position and the **range** switch is set to the desired scale. The probe switch should be in the **normal** position. The rms voltages are read from the two black scales marked "ACV (RMS)," and peak-to-peak voltages are read from the two red scales marked "ACV (P-P)." The same scale multiplication factors as used with the dc voltage measurements are used in the ac mode.

Measuring Alternating Current

Although no direct-reading alternating current scales are included in the FE20, the meter can be used to measure ac using the adapter circuit shown in Fig. 2-3. The appropriate resistors are placed in series with the ac source and the device in which current is to be measured. The FE20 is set up for ac voltage measurement and the voltage drop across the adapter resistor is measured. The voltage reading is converted directly to ac amperes. When measuring from zero to 1A with the adapter, the 1-ohm resistor is used and the **range** switch is in the **1V** position. To measure 3A of alternating current, the switch is closed so that there are three 1-ohm resistors in parallel. A current of 3A will now produce a 1V deflection, and the 0-3.0 scale can be read for 3A.

Measuring Resistance

To make conventional VOM-type resistance measurements, the FE20 **function** switch is placed in the **hi power ohms** position and the **range** switch is set to the range that will put the resistance to be measured as close to center scale as possible. When the **function** switch is set for the ohms functions, the meter pointer will move from the left edge of the meter (zero) to the infinity mark on the right side. Use the **ohms adj** control to adjust the meter pointer to the infinity mark while the leads are held open. When the meter leads are shorted together, the meter will indicate zero ohms on the left side of the meter. Be sure that the probe switch is in the normal position. Any resistance on the Rx1 range is the resistance of the switch contacts and internal wiring in the meter circuit.

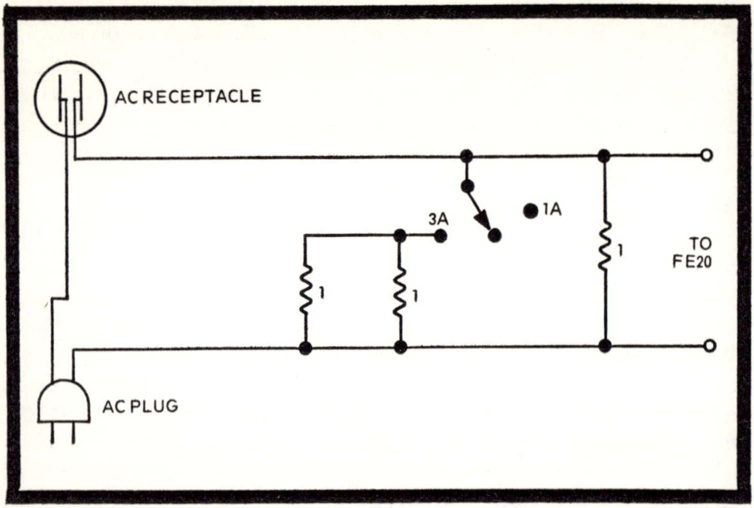

Fig. 2-3. Schematic of ac adapter circuit for the Sencore FE20. (Courtesy Sencore.)

This small value of resistance (approximately 1.4 ohms) should be subtracted from the measured reading.

When using the **hi power ohms** function in solid-state circuits, the applied voltage from the FE20 will be high enough to cause the diodes or transistors to conduct. This can be used to advantage, because the front-to-back ratio of the device may be measured by reversing the meter leads.

The **lo power ohms** function is used to make resistance measurement in solid-state cirucits whn the resistance caused by the conducting diode or transistor junction would cause objectionable error. The resistance measured with the FE20 in this mode will be the resistance of the circuit components surrounding the transistor or diode, and will not be affected by the conduction of the device, since the **lo power ohms** function applies only 0.08V to the circuit under test. Operation in this mode is the same as in the conventional mode.

Output Power Measurements

The output power of a circuit is often measured in decibels (dB) with no reference to the actual power in watts being made. The decibel value is a ratio of power output to power input, calculated as follows:

$$dB = 10 \log \frac{P_{out}}{P_{in}}$$

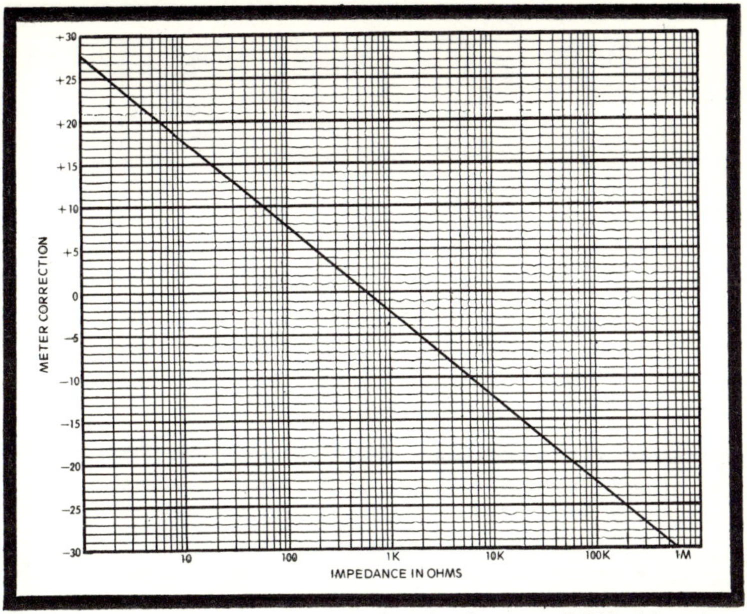

Fig. 2-4. Decibel correction scale for the Sencore FE20. (Courtesy Sencore.)

where **P** is power in watts, or

$$dB = 20 \log \frac{E_{IN}}{E_{OUT}}$$

where **E** is voltage in volts, or

$$dB = 20 \log \frac{I_{IN}}{I_{OUT}}$$

where **I** is current in amperes.

Although the power formula is independent of the input and output impedances, the voltage and current formulas will only be true when the input and output impedances are the same. In many cases, the input and output impedances of the amplifier differ.

In order to use decibel values correctly then, input and output conditions must be specified. It is generally accepted that 0 dB is equivalent to 6 mW (0.006W) of power dissipated in a 500-ohm impedance. Another decibel reference is the **dBm**, or 1 mW (0.001W) into a 600-ohm impedance. This is the reference level used with the FE20 for decibel measurements.

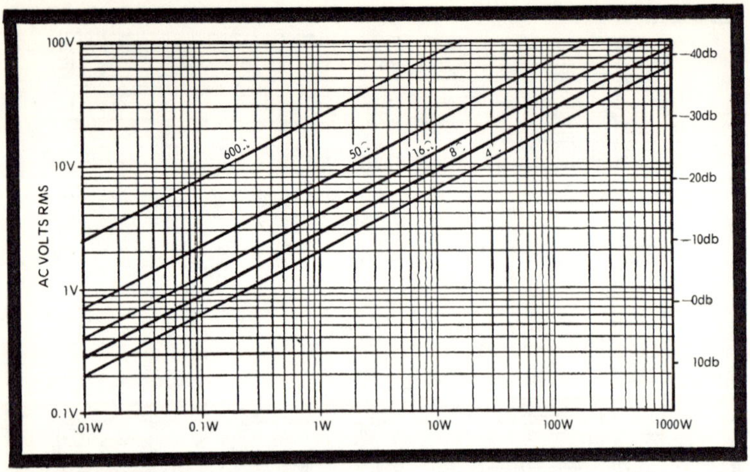

Fig. 2-5. Voltage-vs-power chart. (Courtesy Sencore.)

The nine ac voltage ranges of the FE20 are in 10 dB steps, so that starting from 0 dB on the 1.0V range, you add 10 dB for each range above 1.0V and subtract 10 dB for each range below 1.0V. If the impedance across which the decibel measurement is being taken is other than 600 ohms, it will be necessary to use the chart in Fig. 2-4 to determine the actual decibel level. For example, assume that the meter reading is —4 dB with the range switch on the 100V scale, and the impedance of the circuit under test is only 6 ohms. The 100V range is four ranges higher than the 1.0V range, so adding 40 dB to the meter reading results in 36 dB. Now, referring to the chart in Fig. 2-4, the correction factor is +20 dB, for an actual measurement of 56 dB.

The chart in Fig. 2-5 is used to determine the actual power dissipated in a circuit if the impedance and either the applied voltage or the decibel level (as measured with the FE20 meter) is known. The chart also shows the relationship of ac volts to decibels. The decibel scales on the FE20 and the charts in Figs. 2-4 and 2-5 only apply when the voltage being measured is a pure sine wave. Since the FE20 is a peak-to-peak meter, the complex waveform of music or speech will not give a true rms reading for the decibel-measuring system to work with.

3

Troubleshooting With The EVM

Troubleshooting electronic equipment is a process of reasoning. This is true regardless of the type of equipment being tested or the instrument being used to observe circuit performance. In any troubleshooting situation, the technician must make certain observations of circuit performance, and based on the knowledge gained from those observations, either determine the faulty component or decide what other tests are necessary.

Two of the most valuable troubleshooting aids in this process are a block diagram of the equipment under test and a thorough knowledge of the circuit in question. Space does not permit the explanation of all of the different types of circuits encountered in the electronics service business. However, one can take a very simple block diagram and illustrate the idea behind a logical approach to troubleshooting. Keep in mind that at this point we are interested in troubleshooting using the EVM.

A LOGICAL APPROACH TO TROUBLESHOOTING

Figure 3-1 is a simplified block diagram of an AM-FM radio. With this receiver, one can demonstrate four important points of logical troubleshooting: common connection, points of convergence, points of divergence, and "halve and try."

Common Connection

The idea of **common connection** can be very simply illustrated by the old tube-type radios that employed parallel filaments. If all of the filaments were out, obviously the trouble was somewhere in the power supply, where it affected all stages. If only one filament were out, then the trouble was pinpointed to a stage.

When troubleshooting a circuit, this is one of the first things to look for: What is the overall trouble? In Fig. 3-1, for example, one might ask, Does the fault affect both AM and FM operation, or just one of them?...Is the receiver completely dead?...What are the overall symptoms?" After a thorough

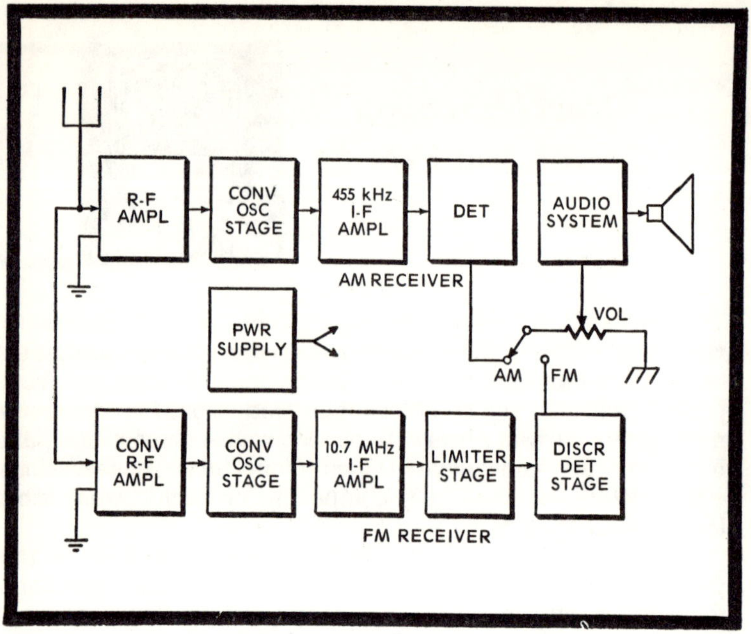

Fig. 3-1. AM-FM receiver block diagram.

investigation has revealed all the faults, ask yourself, "Is there any common connection between the various symptoms?"

Looking at Fig. 3-1 we can see that three areas in the block diagram can produce troubles that will be common to both AM and FM operation. The audio section, from the volume control through the speaker, can produce troubles which generate symptoms in both modes. So can the antenna in this particular set. Since the power supply furnishes the necessary supply voltage to all stages, troubles here can also affect both modes. From this it should be seen that any trouble present in both modes of operation probably originates in one of these common areas. Of course, further testing will be necessary to see just which of these areas contains the fault.

Convergence and Divergence

This is where the **point-of-convergence** concept comes in. Notice the AM and FM functions share the same path from the volume control through the speaker. The volume control, then, is a point of convergence, since the AM and FM signal paths converge at this point. If the receiver is completely dead or if the fault affects both AM and FM operations, it would be wise to begin testing at the point of convergence.

A similar idea can be seen in the **point-of-divergence** concept. In Fig. 3-1, the AM and FM signals separate at the rf inputs. If the antenna is at fault, signals fed to both the AM and FM chains will be out of tolerance. The more complex the piece of equipment, the more important the points of divergence, convergence, and common connection become.

Halve-and-Try Method

Once it has been established that the fault lies in a particular signal path, we come to another troubleshooting technique, "halve and try." Let's say, for example, that the trouble lies in the audio section of the block diagram, somewhere between the volume control and the speaker. To illustrate the point, assume that an audio signal is present at the wiper arm of the volume control but no output is available at the output of the final stage. Where is the most effective point to make the next test? The best place is between the second amplifier and driver stage. This is halfway through the audio chain and, depending on the results of the test, will isolate the trouble to either the driver or output stage, or to the two amplifiers. Only one other test is necessary to locate the faulty stage.

Notice that in this concept the next test point will always be halfway between the point of normal operation and the point where no signal is being obtained. The advantages of this halve-and-try method can be demonstrated by Fig. 3-2. Assume that this diagram is for an electronic device containing eight stages, and the fault is such that there is an input to stage 1 but no output at point H. If you went through and checked the output of each stage from stage 8 back to stage 1, it would take a minimum of 7 tests to isolate a first-

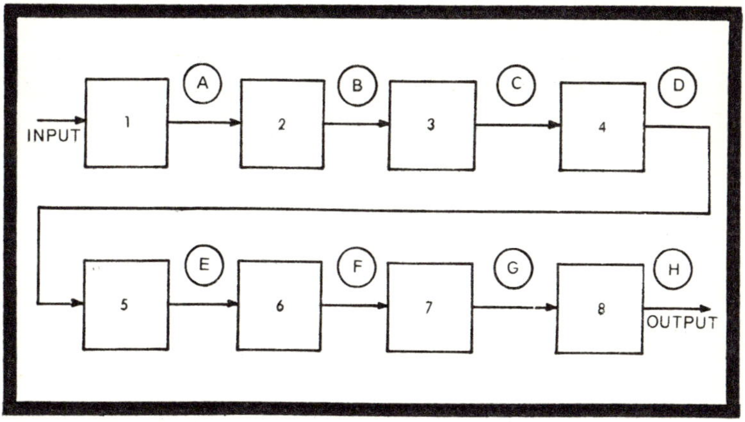

Fig. 3-2. Illustration of divergence and convergence.

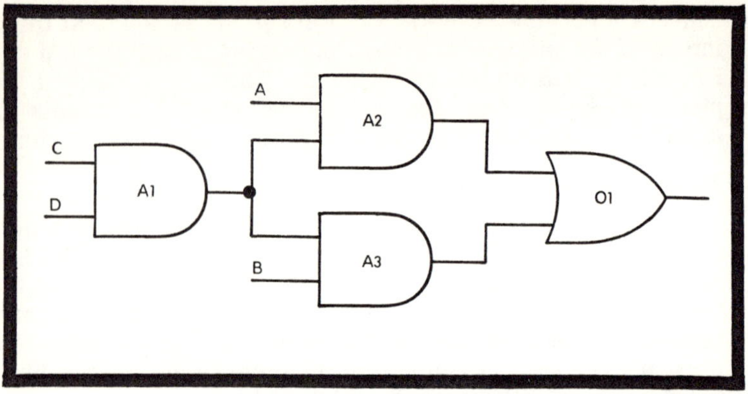

Fig. 3-3. Stage isolation techniques.

stage trouble. If you worked from the other end, it would take the same number of steps to locate an eighth-stage fault.

Let's look at a better way. With the same conditions as before, we make our first test at point D. This is halfway through the signal channel. If the signal is good at this point, we have isolated the trouble to one of the last four stages. Should the signal be missing here, then the trouble is in one of the first four stages. Assuming that there is a good signal at point D you would try point F next. This again divides the trouble area in half. After F, only one test will be necessary to locate the faulty stage. If the signal is good at D and F, then a check at G will determine the bad stage.

The divergence and convergence concepts of troubleshooting are frequently used by technicians servicing digital computers in which there is a tremendous complex of interfacing circuits. A computer logic block diagram in Fig. 3-3 is used for illustration. The complete circuit consists of four gating stages: three AND gates and one OR gate. For the purpose of this explanation, it will suffice to say that the AND gates will generate an output when, and only when, both of its inputs are present; and the OR gate will generate an output when either or both of its inputs are present. The AND gate, A1, receives two inputs, labeled "C" and "D." The C and D lines, then, converge at gate A1. The output of A1 is routed to two circuits, gates A2 and A3; therefore, gate A1 is also a point of **divergence**. At gate A2, the output of gate 1 converges with a third input labeled "A" and, at gate A3, the output of gate 1 is shown converging with the fourth input labeled "B." The OR gate, O1, is a point of convergence, since the outputs of gates A2 and A3 meet there. The function of gate O1 is to generate an output when either gate A2 or A3 has an output. Gate A2 will have an output if there is an output from gate A1 and there is an input on the line labeled "A." Gate A3 will have an input whenever gate A1 has an output and there is an input on the line labeled "B." Gate A1 will generate an output when inputs C and D are present at the same time.

Point-of-Convergence Concept

Now let's illustrate the point-of-convergence concept. If the output of gate O1 is present each time it should be, then we can assume that all circuits ahead of gate O1, and the inputs that feed these circuits, are operating properly. But suppose the output of O1 is not present when it should be. Then we have to investigate a little further. Is the output of O1 ever present, or does it fail only part of the time? If there is never an output, then we can logically eliminate gates A2 and A3 and their associated inputs as the cause. We do so because it would take a simultaneous failure of both A2 and A3, or inputs A and B, to cause a complete loss of output. This is not likely to happen, at least not as often as the failure of a single gate—A1, for example. Since the diverging output of gate A1 is routed to both A2 and A3, the most likely spot to locate this trouble would be in gate A1 or one of its inputs. Gate O1 itself could also be the problem circuit.

Suppose that the failure is not constant; sometimes there is an output from gate O1 and sometimes the output is missing. Then a check of the inputs to O1 would be in order. Since signals converge at O1, we must locate which input contains the fault. Gates O1 and A1 are not good suspects, since a failure in these gates would cause a complete loss of output. Then, either gate A2 or A3, or one of their associated inputs, must contain the fault.

Now let's go back to the broadcast receiver in Fig. 3-1 and apply this same method of troubleshooting to a more conventional circuit. Notice the similarity of the simplified diagram in Fig. 3-1B and the logic diagram in Fig. 3-3. In the audio section of the receiver we have three inputs converging: the power supply output, the AM output, and the FM output. To have an audio output, the power supply output and the output of either the AM or FM section must be present. Therefore, if the failure is present in both the AM and FM modes of operation, the power supply is a likely spot in which to look for trouble. However, two other common (or convergence) circuits could be at fault.

Notice that the antenna has a diverging output going to both the AM and FM sections. Also, don't overlook the possibility of trouble in the audio section itself. The audio section can be eliminated by checking for an input at the point of signal convergence, the volume control in this instance. If a signal is present here, the audio section can be blamed; if not, we are down to two circuits, the antenna and the power supply. A quick check of power supply voltages would give the answer.

Using this procedure in Fig. 3-2, you can see that once it has been established that a good input is present and no output exists at point H, then a maximum of three tests are necessary to isolate the faulty stage. In the stage-by-stage testing technique, it could take up to seven tests to isolate some problems in Fig. 3-2.

STAGE TESTING

Once the faulty stage has been isolated, other tests are in order to pinpoint the trouble within that stage. The exact method used to test an individual stage depends on the type of circuit (amplifier, oscillator, etc.). In many cases, proper testing with an EVM can confirm the suspicion formed during the stage isolation phase of troubleshooting. The tests made with the EVM are usually ac and dc voltage measurements within the stage.

Most electronic circuits have two different types of current paths, as shown in Fig. 3-4. A dc circuit exists within the stage to establish the supply and bias voltages, thus determining the operating characteristics of the stage. In the figure, the dc circuit has been redrawn in B, omitting the ac parts. The collector voltage is furnished by the power supply through resistor R3. Resistors R1 and R2 form a voltage divider that furnishes base bias for the transistor. With forward bias present, the transistor will conduct, and dc voltages will be present at various points throughout the circuit. In addition to current from the negative terminal of the battery through R2 and R1, there will be a path from the negative terminal of the battery, through R4 to the collector of Q1, through the transistor and R3, and back to the positive terminal. If the values of R1, R2, and R3 are such that reverse bias is achieved, the transistor is cut off and no current exists in R3 or R4.

The ac circuit is illustrated in Fig. 3-4C. Here the components that handle the various alternating currents are shown. Assuming that an ac signal is present at the input, C1 couples this signal across R2 to the base of the transistor. As the base is driven alternately in a positive and negative direction, current through the transistor increases and decreases in step with the signal, causing alternating currents in the collector and emitter circuits. Capacitor C2 shunts alternating current around R4, thus preventing any ac voltage from being "felt" at the emitter of Q1. Capacitor C3 couples the alternating output at the collector to R5. Should any large swings in collector voltage be "felt" through R3, C4 routes them to ground. Therefore, we have established the principle ac paths indicated by the arrows.

Should anything disturb the value of the components of this circuit, the current paths will be altered. This will result in

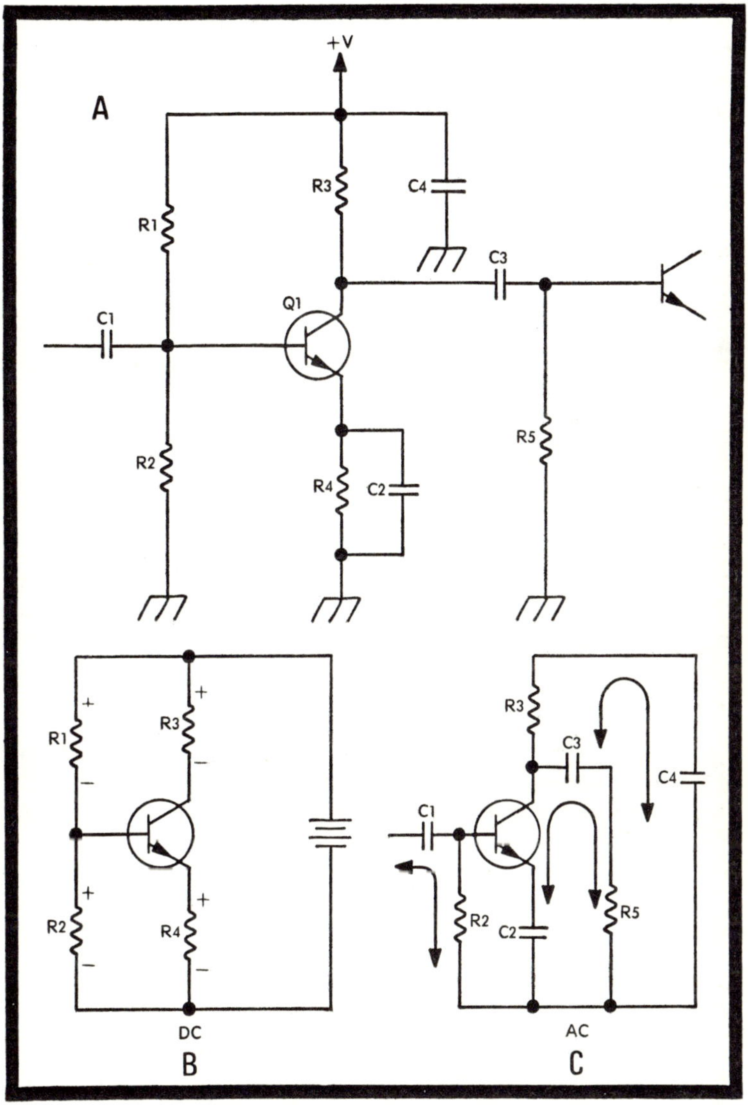

Fig. 3-4. Ac and dc paths in a typical amplifier.

abnormal voltages—either ac or dc, or both—being present in the circuit. By means of the EVM we can test the operation of the various electronic circuits.

The ac voltages can be measured with an EVM, using the proper ac voltage range. Keep in mind that the frequency of the signal should be within the response limits of the meter, for readings to be accurate.

The dc voltages can reveal a lot about the circuit. If base voltage is normal and emitter voltage is zero, obviously the transistor is not conducting. If the collector voltage is low and the emitter voltage high in this particular example, the transistor is conducting too heavily.

Let's see how we can apply this to actual circuit testing.

Amplifiers

We will start with an RC-coupled amplifier to demonstrate how the EVM can be used to determine whether an amplifier is working properly and to show some of the basic steps in locating faults within the amplifier. Again we will use Fig. 3-4 for our example.

To test this circuit, an alternating input such as that available from a signal generator is needed at C1. An EVM set for ac operation is placed across output load resistor R5. If the input signal is within the frequency response range of the amplifier, an ac voltage is developed across R5. This is the quickest way to check overall operation using an EVM. If there is no output across R5, move the meter (still set for ac voltage) to the collector of the transistor and then to the base. This will check for an open C3 or C1. You are chasing the signal through the stage just as we chased it between stages in the previous examples. Keep in mind that only a small ac voltage should be present at the emitter, due to C2.

If ac voltage measurements fail to reveal any clues as to the trouble, shift the meter to the proper dc voltage scale and measure the voltages at the terminals of the transistor. Too high a collector voltage points to underconduction, and too little collector voltage points to overconduction. The latter condition is confirmed by a simultaneous excessive emitter voltage. Once the conduction level of the transistor is determined, look for causes of these conditions. The causes also can be determined by voltage checks sometimes.

Tuned Amplifiers

A tuned amplifier can be tested in much the same manner as the amplifiers already discussed. A stable signal at the amplifier's operating frequency must be applied to the input. If the circuit does not already include a detector, one such as that shown in Fig. 3-5 can be used in the meter probes. The detector rectifies the output and supplies a test voltage within the frequency limits of the EVM.

To set up the test, connect a tone-modulated (400-1000 Hz) signal to the amplifier's input, set the EVM to the proper ac scale, and connect it to the detector circuit. A quick test can now be made of circuit operation by rocking the signal generator **frequency** control back and forth past the amplifier operating frequency. The output, as indicated by the EVM, should increase as the operating frequency is approached and fall off as the signal generator output is moved off of the amplifier frequency.

Frequency response is a measure of an amplifier's ability to amplify various frequencies. Depending on whether the amplifier handles audio or rf frequencies, either the setup described for Fig. 3-4 or for Fig. 3-5 may be used. The purpose of this measurement is to plot the amplifier's output level at various input frequencies. As the input is stepped from frequency to frequency, the output level is measured and recorded. The result will be an indication of the amplifier's ability to amplify various frequencies, and thus an indication of its operating condition.

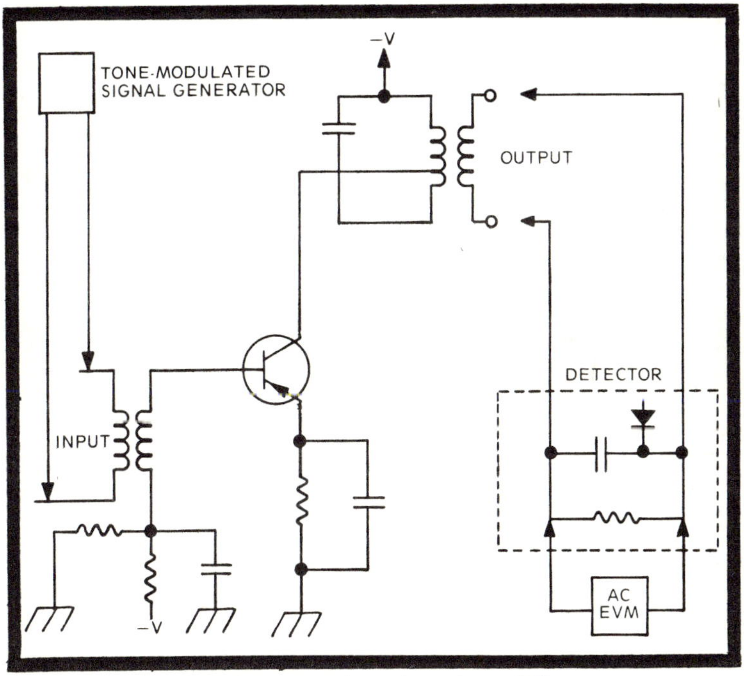

Fig. 3-5. Signal testing a tuned amplifier.

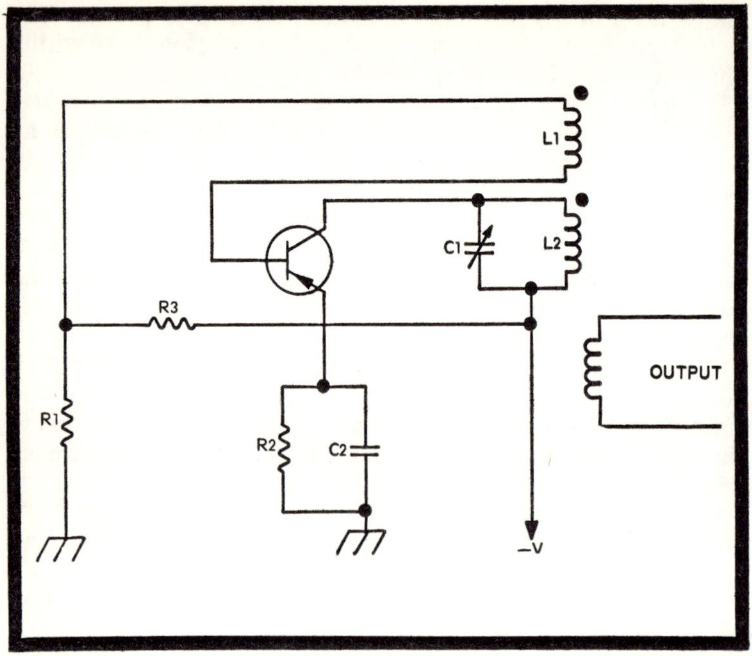

Fig. 3-6. Typical oscillator.

In some instances, it is convenient to use decibel measurements for this type of testing. If decibel measurements are to be used, the input-output relationship is stated with no reference to any particular power or resistance. To use the EVM for such measurements, set the input signal to the specific frequency at which the tests are made. Now adjust the amplitude of the signal input to 0 dB on the EVM. Move the EVM to the output and make a second decibel measurement. This gives the gain of the amplifier in decibels. The procedure is repeated for each frequency at which measurements are desired, giving a useful indication of the amplifier's frequency response.

Oscillators

In older equipment using vacuum tubes, oscillators were checked by measuring the grid bias. Most of these circuits employed grid-leak bias; therefore, a negative potential was present on the grid when the oscillator was operating properly. The absence of this negative potential was an indication that oscillations had ceased.

This same principle can be employed in solid-state circuits if a few points are kept in mind. Remember that a transistor must be forward biased in order to conduct; therefore, there must be some bias present at all times. In this case, then, we must look for a **change** in oscillator bias. With the EVM set to the proper scale, monitor the base bias voltage as the oscillator tuning capacitor is shorted out. Do not use a direct dc short for this, since it might cause heavy dc currents and damage some circuit component. Instead, use a capacitor of about 0.01 uF to parallel the tuning capacitor (C1 in Fig. 3-6). If no change in bias voltage is noticed, the oscillator is inoperative. Base bias for the oscillator in Fig. 3-6 is supplied by voltage divider R1 and R3. Positive feedback, necessary to sustain oscillation, is provided by L1.

COMPONENT TESTING

Besides being used for stage testing, the EVM can also be used to test individual circuit components, both in and out of their circuits. In some cases, these tests are not completely conclusive; however, they do offer a convenient way to make fast tests that will usually be accurate enough for troubleshooting purposes.

Resistors

An unknown resistance may be measured by setting the EVM up for resistance measurements and connecting the unknown resistance between the probes. The most accurate measurements are obtained when there is no parallel path for current around the unknown resistance.

In some cases, reliable in-circuit resistance measurements can be obtained using an EVM equipped with a low test voltage feature (such as the **lo power ohms** mode of the Sencore FE20). This mode of operation employs an extremely sensitive amplifier to amplify the current generated through the unknown resistance. Features such as this enable the EVM to be used for in-circuit resistance measurements without the need for disconnecting the solid-state components connected to the resistance under test. For example, suppose you want to measure the resistance of R1 in the power supply circuit shown in Fig. 3-7. Using a conventional ohmmeter across R1, as shown in B, may forward bias rectifier X1. Current then flows through the rectifier, the transformer secondary, and the load (R_L), giving an erroneous reading. To obtain an accurate reading, the resistor would have to be disconnected at one end before making the measurement (Fig. 3-7C). By using a test voltage that is below the forward bias voltage of X1, an EVM can be connected as shown in B, and accurate readings still can be achieved.

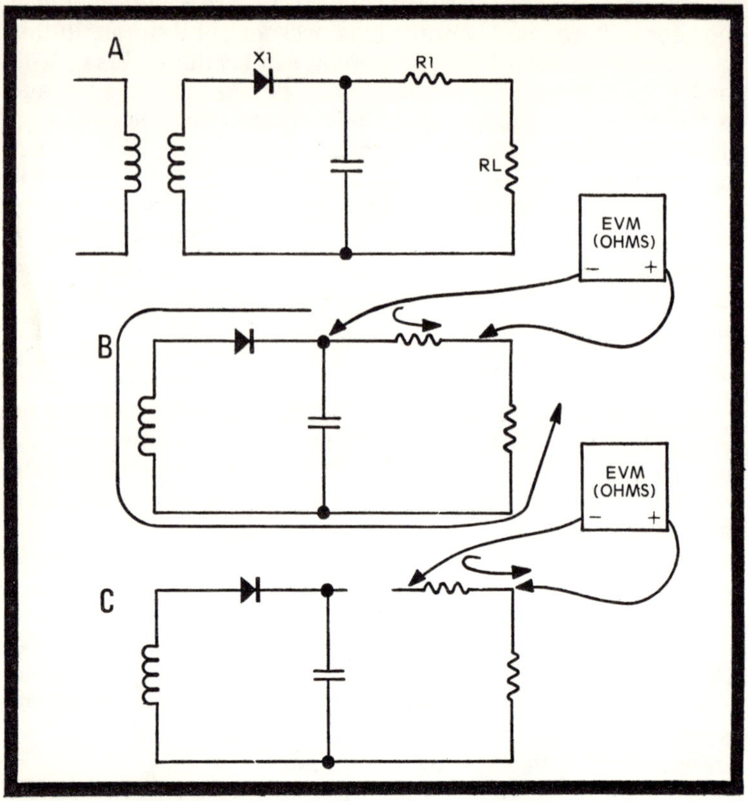

Fig. 3-7. In-circuit resistance measurements.

Capacitors

Although an EVM cannot be used to measure capacitance directly, it can be used in some cases to give a general indication as to a capacitor's condition. Some typical examples of in-circuit capacitor testing are illustrated in Fig. 3-8.

In A, the emitter bypass capacitor is being tested using the ac voltage function of an EVM. The purpose of this capacitor is to bypass the ac signal in the emitter circuit to ground. By measuring the ac voltage across the capacitor, we can determine just how well the bypass function is being performed. Above-normal ac at the emitter means the capacitor is not performing properly. Figure 3-8B illustrates a method of testing coupling capacitors. Under normal operating conditions, the capacitor couples the ac input across the base resistor. By measuring the ac voltage drop across the resistor we can get a rough idea of how well the capacitor is doing its job.

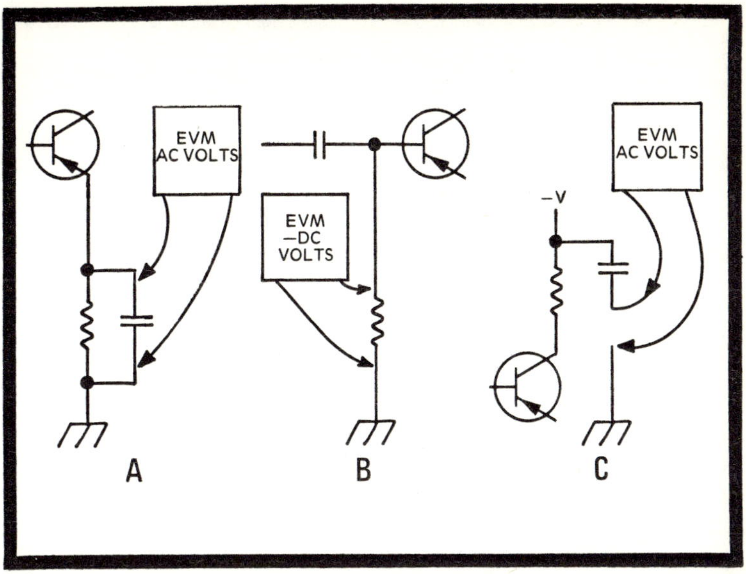

Fig. 3-8. In-circuit capacitor testing.

One function of a capacitor, of course, is to block dc. Figure 3-8C shows a capacitor being tested for its ability to do that. The capacitor is connected at one end to the power supply. An EVM set for dc voltage measurement is connected to the opposite capacitor lead. The lower the dc voltage, the better the capacitor is blocking the direct current. Conversely, excessive dc voltage means the capacitor is leaky.

An indication of leakage resistance can sometimes be obtained using the test setup in Fig. 3-9. Here the EVM is set up for resistance measurements. With nothing connected between the probes, the meter indication will be infinity. As soon as the probes are connected to the capacitor, a charging current in the capacitor will cause the pointer to swing toward zero ohms. Then, as charging current decreases, the pointer slowly falls back toward infinity, coming to rest at some value in between.

All of these tests are "rough checks." They will not give a perfect indication of the capacitor's condition, but usually they are good enough to determine if the capacitor is operational. For example, a capacitor that is only slightly leaky can cause an oscillator to drift completely off frequency. The test described here may not reveal the leakage unless it is quite large. On the other hand, this type of testing will give a rapid indication as to whether the capacitor is shorted, open, or excessively leaky.

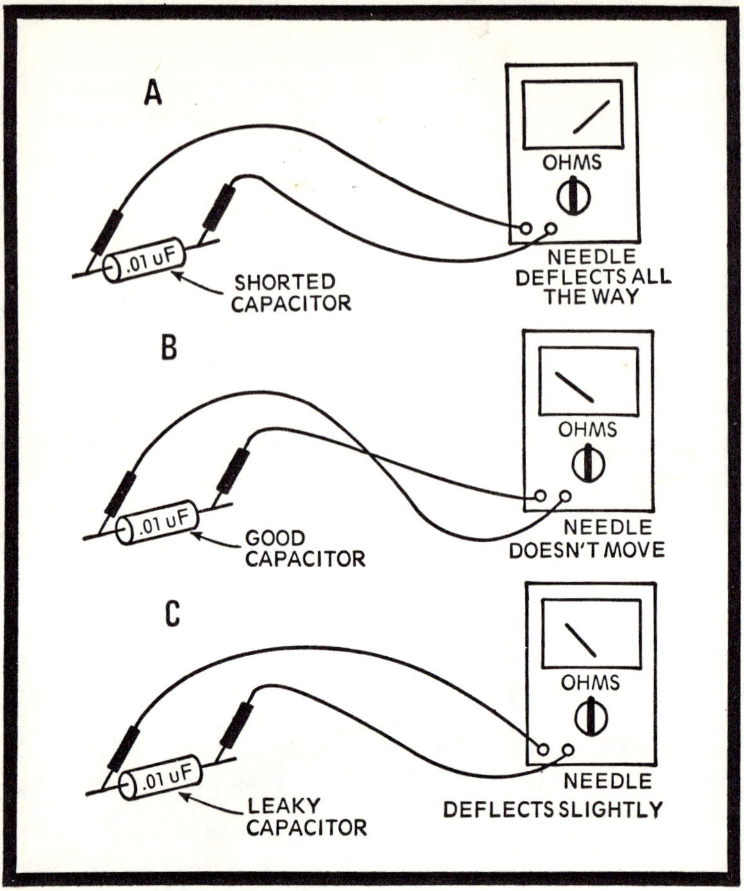

Fig. 3-9. Out-of-circuit capacitor testing.

Coils, Chokes, and Transformers

The windings of inductive devices are subject to four types of failures: opens, shorts between the windings, shorts between turns of the same winding, and shorts between a winding and the core. In-circuit voltage tests are helpful in isolating some of these problems, and a continuity test can be used to detect opens. Inductors cannot be completely evaluated with an EVM, but troubleshooting tests can be made.

The easiest test is the continuity check. This test will reveal any open windings. Set up the EVM to measure small resistances, and measure the resistance of each winding of the inductor. In-circuit voltage tests can sometimes achieve the same result, as shown in Fig. 3-10.

Shorts between the individual windings and between a winding and the core can be located using the EVM as an ohmmeter. All coils should be separated by a high resistance unless there is an internal connection. Check the schematic to be sure. The resistance between the core and any coil should be high—in the megohms region.

Shorts between windings of the same coil are the hardest to locate using an EVM. If the coil is in a tuned circuit, this can sometimes be done by monitoring the output with a meter while the circuit is being tuned, noting how the circuit responds to tuning procedures. Shorts in transformer windings will sometimes cause excessive heating, particularly near the short.

Batteries

The battery is a frequent cause of poor operation in portable electronic equipment. Unfortunately, it is also a component that is frequently subject to poor testing practices. In many cases, the battery will appear to be in good shape if removed from the circuit. For this reason, batteries should be in the circuit and under normal loading conditions when tested. Measure the battery voltage while the battery supplies normal circuit current.

Don't overlook poor connections at the battery terminals as a cause of low circuit voltage. Even though the battery may be in good condition, high resistance connections can impede current flow to a point where circuit operation is degraded. The best way to test for high resistance in a connection is to measure the voltage drop across the connection.

Meter Polarity

Before testing solid-state components with the EVM ohmmeter, the polarity of the meter's test voltage must be

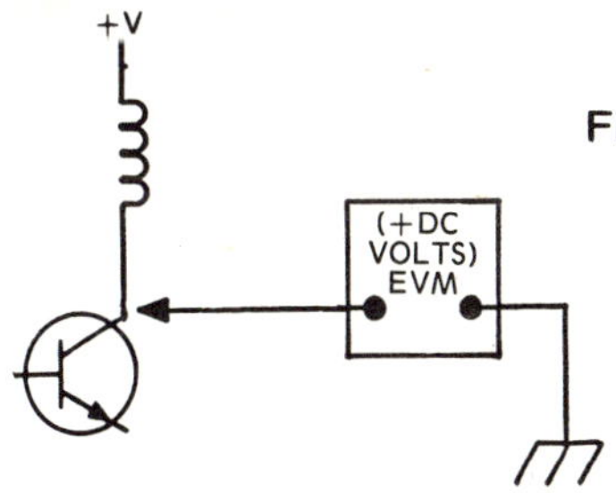

Fig. 3-10. In-circuit coil testing.

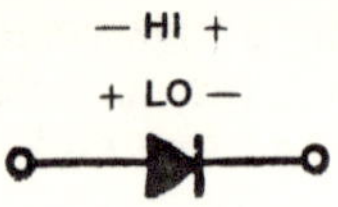

Fig. 3-11. Diode testing.

determined. This can be done with a dc voltmeter or a diode known to be in good condition. If a dc voltmeter is used, the negative lead is connected to the one lead of the EVM and the positive lead is connected to the other. If the dc voltmeter reads a positive voltage, then the EVM lead connected to the positive dc voltmeter lead is the positive terminal of the EVM ohmmeter circuit. Should the dc voltmeter read a negative voltage, the EVM lead connected to the dc voltmeter's positive lead is negative.

To establish meter lead polarity using a diode, connect the meter leads to the diode terminals and note the resistance reading. If a low resistance is obtained, the diode is forward biased, indicating that the negative lead is connected to the cathode and the positive lead to the anode. A high resistance indicates opposite polarities.

Diodes

Solid-state diodes may be tested fairly accurately either in or out of a circuit by using an EVM to measure the **front-to-back ratio**. (Recall that a forward-biased diode has a very low resistance.) The EVM is set for ohmmeter operation; usually the Rx100 scale gives the best results. When the meter probes are placed across the diode, a low resistance should be indicated in one direction (when the diode is forward biased) and a high resistance in the other direction (see Fig. 3-11). The higher the ratio between the front and back resistances, the more effective the diode. If the readings are very high (indicating an open) or both readings low (indicating a short), the diode must be discarded.

When testing diodes using the EVM, make sure that the test voltage applied to the diode does not exceed the voltage and current ratings of the diode under test.

Transistors

Several tests of a transistor can be made using the EVM. The most popular of these is the **junction** test. In the procedure given here, an npn transistor is used; however, the same setup may be used for a pnp type, except the polarity of the ohmmeter probes must be reversed.

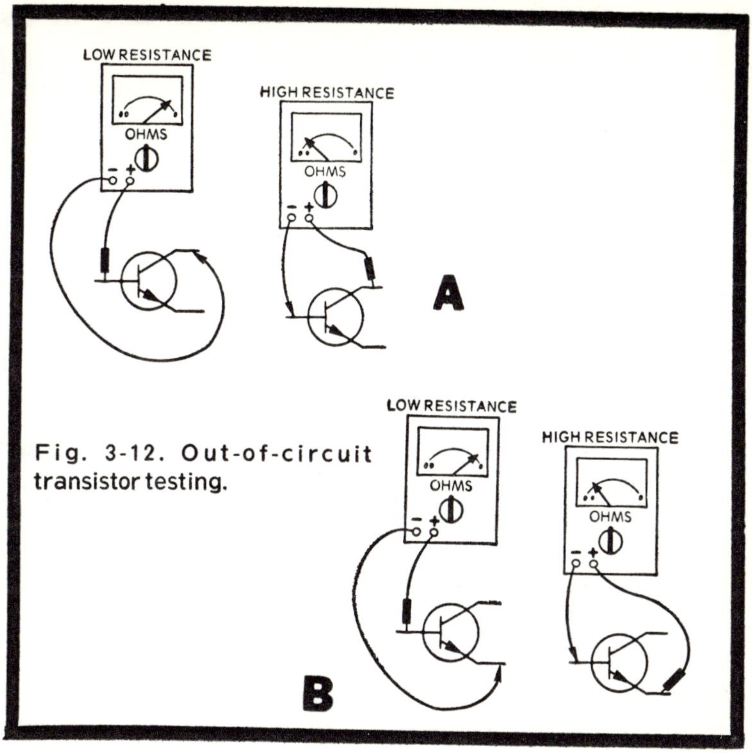

Fig. 3-12. Out-of-circuit transistor testing.

For low-power and signal transistors, the EVM is set to the Rx100 scale, while high-power transistors require a lower setting—either Rx10 or Rx1. Place the negative lead of the EVM to the emitter and the positive lead to the base, as shown in Fig. 3-12. The meter should indicate a low resistance. Now reverse the leads; the meter should indicate a high resistance. The actual value of the readings will depend upon the type of transistor and the particular EVM being used. The important thing is the ratio between the two readings. If they are nearly the same, the transistor is defective.

Now make a similar test of the base-to-collector junction, as shown in Fig. 3-12B. With the negative probe at the base, a low reading should be obtained. Finally, measure the emitter-to-collector resistance. This should be relatively high in both directions; however, a lower reading is obtained when the negative lead is connected to the emitter. Another simple test can be made without disconnecting the transistor from the circuit. In the circuit shown in Fig. 3-13, R1 and R2 supply base bias for the transistor. Normally, this places forward bias on the transistor, and current flows. Collector voltage is

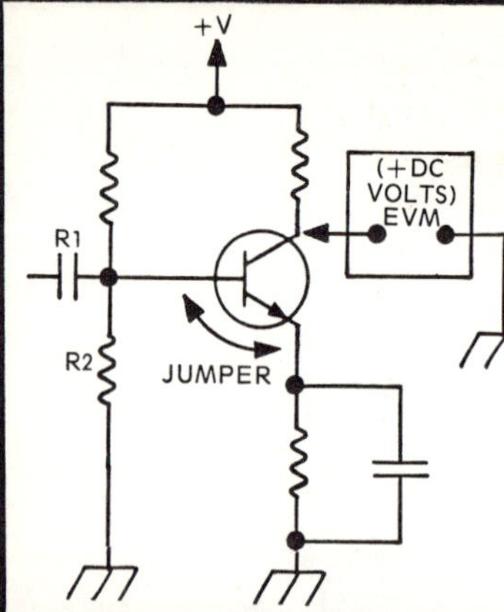

Fig. 3-13. In-circuit transistor testing.

somewhere below the supply voltage. By means of a shorting jumper, the emitter and base are placed at the same potential; the transistor should cut off, causing collector voltage to approach supply voltage.

FETs

Before troubleshooting tests can be made on an FET with an EVM, one must first determine the type of FET (JFET or MOSFET). The JFET, as shown in Fig. 3-14A, has its gate electrically connected to the source-drain channel; thus, a junction surface similar to one in a bipolar transistor exists and may be tested for resistance.

A forward resistance check can be made using the low-voltage-ohmmeter feature of the EVM, with the range control set to Rx100. For n-channel FETs, connect the negative lead of the meter to the gate and the positive lead to either the source or drain. The reading should be low. (For p-channel devices, the opposite connections are used, as shown.) Now reverse the

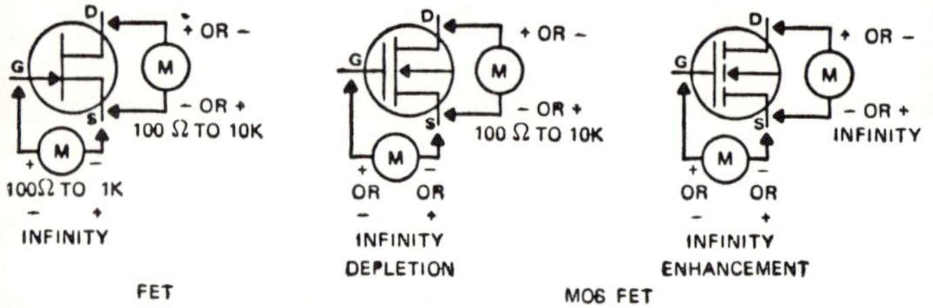

Fig. 3-14. Testing an FET.

meter leads and look for an almost infinite resistance when the junction is reverse biased.

Testing an insulated-gate FET (IGFET or MOSFET) with an EVM is somewhat limited. (Refer to Fig. 3-14A and B.) The same resistance readings made with the JFET apply; however, since the MOSFET has its gate insulated from the channel, the reading should be extremely high in both directions. Use the highest resistance scale available in the low-test-voltage mode, for best results.

SCRs

To test an SCR for its "firing" ability, connect the EVM as shown in Fig. 3-15, using the Rx1 scale. A high resistance reading should be obtained. Now use a jumper to short the gate to the anode. This should cause the gate to fire and the resistance to drop to a value somewhere under 50 ohms. Remove the short and notice that the resistance reading does not change. The reading should remain the same until the SCR is cut off. Remove the meter leads from the anode and cathode. Now, when the leads are again touched to the SCR, the resistance should remain high until the gate is again shorted to the anode.

Zener Diodes

A quick test of a zener diode may be made in a manner similar to that for the conventional diode. This test checks the rectification of the zener, but provides no information regarding its regulating properties. As with other tests, however, it is useful for troubleshooting.

Final Note

Each of the devices discussed here presents different testing problems. For that reason, the EVM alone cannot be

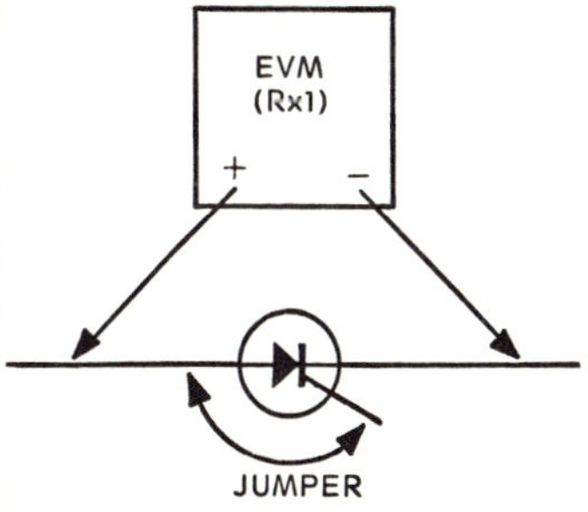

Fig. 3-15. Testing an SCR.

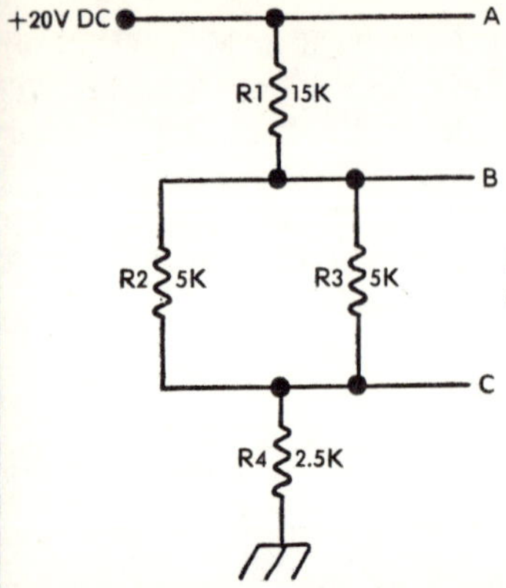

Fig. 3-16. Resistive circuit voltage analysis.

used to perform conclusive tests in each instance. Also, the tests described here are not the only ones that can be used for troubleshooting. Always follow the manufacturer's recommendations for both the EVM and the circuit under test when troubleshooting.

VOLTAGE ANALYSIS

Although it is sometimes necessary to make voltage, current, or resistance measurements of a single component, the most-used method of testing with an EVM is a voltage analysis of the circuit in question. In most cases, voltage tests can be made easily without disconnecting any of the components. From an analysis of the voltage measurements, current and resistance can usually be determined. Even when the exact current through a component or the resistance of that component cannot be determined directly from the voltage readings, the voltage readings can point to a specific component as the probable cause of trouble, eliminating the necessity of testing each individual component.

Resistive Circuits

The simple circuit in Fig. 3-16 can be used to illustrate voltage readings. The purpose of the components and of the circuit itself is not important to this discussion; only the results of the voltage readings are. Under normal operating conditions, 20V is applied across the circuit between point A and ground. Since the total resistance between B and ground is 5K, or one-fourth the total resistance of 20K, +5V should be measured between B and ground. From point C to ground, one should read 2.5V.

Let's assume there's a normal voltage at point A and +20V at B and C. This should be an easy one. Since the voltages at all three points are the same, one of two conditions must exist: Either a short exists between C and A, causing the entire voltage to be dropped across R4, or there is no current through R4. How can this be verified? Remove all power from the circuit and try a quick in-circuit resistance measurement. Start with a resistance measurement across R4. Under normal operating conditions, the resistance between C and ground will be something under 2.5K. The exact reading will depend upon what is in parallel with R4. If R4 is open, meter test current will have to flow through R1 and the parallel combination of R2 and R3, causing a minimum resistance reading of at least 17.5K. Therefore, if the resistance between point C and ground is less than 2.5K, chances are that R4 is good and a short exists from A to C. If, however, the reading at point C is greater than 17K, look for an open R4.

Now let's assume that the output at A is normal, or +20V, but when a reading is made between B and ground, we find approximately 6.6V. What could cause such a reading? Since the applied voltage at point A is normal, the resistance across which the voltage is applied must have changed. But, which resistance? First think about what must have happened to the resistance. The voltage at point B has increased; this means that either the current through the resistors between point B and ground increased, or the resistance itself increased. From just two voltage readings at points A and B, then, we have been able to deduce that either the resistance between A and B decreased, causing a greater portion of the 20V to be dropped between B and ground; or the resistance between B and ground increased, which also would cause an increase in the voltage drop from B to ground. We can rule out an open R4, because this would cause the voltage at point B to reach the applied voltage of 20V.

A third voltage check would be in order at this time. Say that a measurement between point C and ground reveals a voltage of +1.65V. We can make some observations similar to those made for point B. Since the voltage at point B has dropped, either the resistance between C and ground has decreased or the resistance between A and C has increased—both facts in direct opposition of the conclusions made for point B. So what can we interpret from this?

At point B, we are considering the ratio of the resistances between points A and B, and B and ground (R_{AB} / R_{BG}). At point C, we are looking at the ratio of the total resistance between points A and C, and C and ground (R_{AC} / R_{CG}). The high voltage at point B, we said, could be caused by an increase in the resistance between B and ground—in other words, an increase in the value of R2, R3, or R4. At point C we said the low voltage could be caused by an increase in resistance between A and C. Obviously, R1 hasn't increased in value, or the voltage at point B would be low instead of high. An increase in the value of R4 would cause an increase in voltage at point C; therefore, either the combined value of R2 and R3 has increased, or the value of R4 has **decreased**. A decrease in the value of R4 would not only decrease the voltage at point C, but would also result in a decrease in voltage at point B. So we are left with only one logical conclusion, an increase in the value of the resistance between points B and C (an open R2 or R3, perhaps).

Let's look at another one. This time normal voltage is present at point A, but no voltage is present at point B. Since the meter is already set for voltage measurement, make a quick check at C. If no voltage appears here, either R1 is open or a direct short exists between B and ground. This can be checked just as R4 was checked in the first example: Remove power and measure the in-circuit resistance across R1. It should be less than 15K, under normal conditions. If it reads greater than 15K, R1 is probably open.

Now assume a normal voltage at point A, with 20V present at point B, and 10V at point C. It looks as if the full 20V output of the power supply is being dropped across the series-parallel combination of R2, R3, and R4. There must be a short across R1.

Let's try another easy one. There is normal voltage at point A, and point B reads +2.85V. Point C reads zero volts. There's no voltage drop across R4, yet current flows through R2 and R3. So R4 must be shorted. A short between B and C would be similar in effect, except that B and C would read the same voltage.

One point about shorts: Unless you are working with a very well regulated supply, a short—especially one involving a large portion of the load resistance—is going to "load down" the power supply. In other words, it will cause an increase in current and, in most cases, a decrease in power supply output voltage.

Tuned Circuits

To see how an EVM may be used as a troubleshooting aid in tuned circuits, we will use the simplified schematic in Fig. 3-17. Again, the circuit is not used for any specific purpose, but is a convenient way to explain why certain tests are made. Now, a series-tuned circuit offers minimum impedance at the resonant frequency, while a parallel-tuned circuit presents a maximum impedance at the resonant frequency. In this example, L1 and C1, being a series-tuned circuit between the signal line and ground, offer a minimum impedance at a frequency determined by the values of L1 and C1 (resonant frequency). This tends to shunt that specific frequency to ground, while offering a higher impedance to other frequencies and causing them to pass farther down the line. Parts L2 and C2 form a second tuned circuit and, since they are in parallel between the signal line and ground, tend to further develop the resonant frequency established by L2 and C2, by shunting other frequencies to ground. Parts L3 and C3 form another series-tuned circuit—this time in the signal path. This tunes the signal path line to a specific frequency, making it easy for signals of that specific frequency to continue down the line to L4 and C4. This parallel circuit will block its resonant frequency, since it is in the signal path. From these observations, we can say that the overall function of the circuit in Fig. 3-17 is to pass certain specific frequencies while attenuating others. Similar circuits are found in the rf and i-f sections of broadcast receivers, television receivers, and communications equipment.

Although direct voltage reading analysis is somewhat more difficult in this type of circuit than in the resistive example, it is still possible to make accurate assumptions from voltage readings. The schematic in the service data for this type of circuit usually gives the resonant frequency to

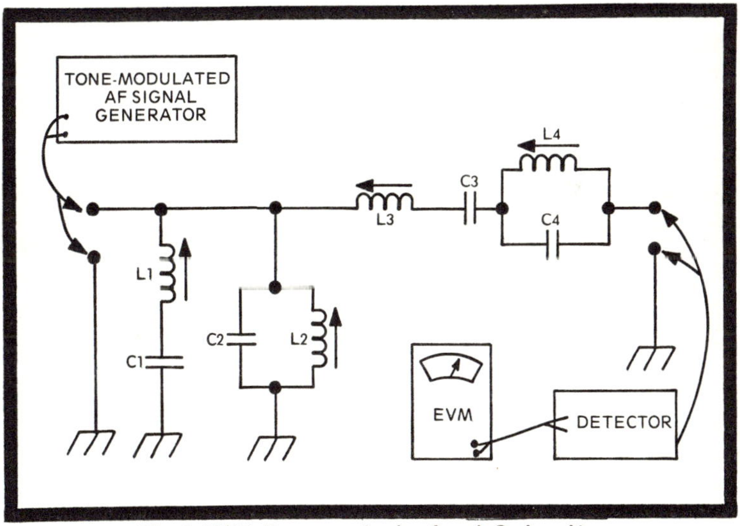

Fig. 3-17. Voltage analysis of an LC circuit.

which each circuit should be tuned. Depending on the frequencies involved and the frequency limitations of the EVM, it may be necessary to use a detector between the test point and meter probe, as shown in Fig. 3-17. In this case, a tone-modulated carrier input is used as the test signal. At lower frequencies, an unmodulated carrier can be used and the detector probe omitted from the meter hookup. The following example will use the tone-modulated hookup shown.

After the meter is connected through a detector probe to the output of the circuit, connect a tone-modulated carrier from a signal generator to the input. Now refer to the manufacturer's service data and note the resonant frequency of each tuned circuit. Starting with the parallel circuit of L4 and C4, tune the signal generator to that specific frequency. The purpose of the L4C4 combination is to block that frequency. Coil L4 is tuned to the setting that gives a minimum output as indicated by the meter. If trouble is suspected, observe the meter reading carefully. As the tuning slug of L4 is slowly turned, the voltage at the circuit output should decrease smoothly until the minimum is reached; then it should increase in an equally smooth manner. Any sudden change or tuning at the wrong frequency indicates trouble. An open or shorted L4 would prevent the circuit from being tuned, and no change would be noted in the output. An open C4 would cause the circuit to be mistuned; however, tuning L4 might cause a change in the output level, but at the wrong frequency.

Next check the operation of the series-tuned circuit of L3 and C3, leaving the signal generator and meter connected as before. Shift the carrier frequency of the generator to the resonant frequency of L3 and C3. This time tune L3 to give a maximum indication on the meter. An open L3 or C3 will prevent an output. If C3 is shorted, there will still be an output, but adjusting L3 will not cause this output to be maximum at the designated frequency. A shorted L3 will not prevent an output, but will detune the circuit so that no effect on the output voltage will be noticed as L3 is adjusted.

The next circuit we encounter as we work our way toward the signal generator is the parallel resonant circuit of L2 and C2. Set the signal generator to the prescribed frequency and tune this circuit for maximum output. Here, a short of either component will prevent any output from the circuit. An open of either component will cause a reduced output at the specified frequency. If C2 opens, the tuning of L2 may still affect the output, but the circuit cannot be tuned. An open L2 will eliminate the effects of any adjustment of the slug.

Parts L1 and C1 comprise a series-tuned circuit to ground and therefore are tuned for a minimum voltage at the output.

Set the generator to the resonant frequency of this circuit and vary the slug of L1, looking for the point of minimum output. An open of L1 or short across L1 would prevent any effect of tuning from being seen. A shorted C1, an open C1, or an open L1 would reduce the amplitude of the output.

Keep in mind when you are working with tuned circuits that severe detuning can eliminate all output. Any partial short to ground (such as in C1) will reduce the overall output, while a dead short directly between the line and ground (in C2, for example) will cause a complete loss of output. An open in a component in the signal path can have the same effect, if it is not paralleled by a second component. For example, an open L3 would cause a loss of output. Shorts between turns of the same coil would cause a detuning effect. In this case, it is sometimes possible to retune the circuit, but not to the desired frequency.

Power Supplies

There are many different types of power supplies found in electronic equipment. Half-wave, full-wave, bridge, and voltage multiplier circuits are some examples. To illustrate how the EVM is used in troubleshooting power supplies, we will use the half-wave rectifier power supply and filter circuit shown in Fig. 3-18. The particular type of power supply is not so important as the troubleshooting techniques presented here.

Two outputs are furnished by this particular supply. A 6.3V ac output is taken from secondary winding L3, and a dc output is taken from filter capacitor C2. Disregarding the effects of the filter circuit consisting of C1, C2, and L4 for the moment, this supply will furnish a pulsating dc to the load

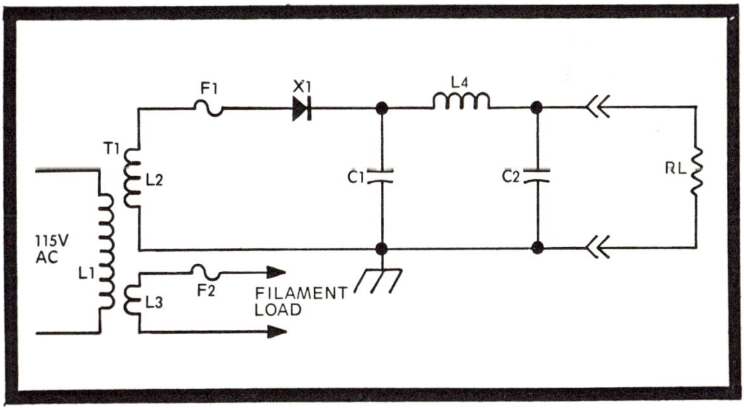

Fig. 3-18. Half-wave power supply.

resistance, R_L. On alternate half-cycles of the ac input, the top of L2 is made positive, forward biasing rectifier X1. On the opposite half-cycles, the top of L2 is made negative, reverse biasing X1 and thus preventing current flow through R_L.

This pulsating dc through the load is smoothed by the action of the filter. When power is first applied to the power supply and the anode of X1 is made positive for the first time, a large initial surge of charging current is drawn by C1, charging the capacitor. (Some charging current is also drawn by C2, but not nearly as much as by C1.) The next time the anode of X1 is made negative, X1 is reverse biased, removing current from the filter circuit and load. Capacitor C1 starts to discharge to ground through C2 and the load, adding to the charge on C2. Before C1 has a chance to completely discharge, the supply voltage polarity is reversed and X1 conducts. Capacitor C1 charges once again. This action continues until C1 and C2 are charged to a maximum. At this time, more of the discharge current and the forward-bias current is drawn by the load. Thus the capacitors tend to keep current flowing constantly, from ground up across the load, smoothing the pulsating dc. Coil L4 aids the capacitors in maintaining a steady current in the load by its characteristic self-inductance, which resists any change in load current.

But what happens when something goes wrong in the power supply of Fig. 3-18? Check the output of both the dc and ac supplies to narrow the fault probability to a specific section of the power supply. If only one supply is bad, then you have localized the trouble to that supply. If both supplies are bad, look for trouble around the transformer. If either output is missing completely, don't overlook the possibility of a blown fuse. Any time a fuse is replaced, the circuit should be checked for shorts before reenergizing the circuit.

A complete absence of a dc voltage across the load with normal ac voltage present at the filament output localizes the problem to the dc supply. Check the fuse by measuring the voltage drop across it. Any value above a few microvolts indicates an open fuse. If the fuse is good, check the ac output of the L2 winding of T1. If a voltage is present and the fuse is good, look for an open X1 or L4. The possibility of a shorted filter or load can be eliminated if the fuse was good. To completely remove all output voltage, a direct short would have to be present, but this would cause F1 to blow. However, it is possible for a leaky capacitor or partial short to reduce the output without blowing the fuse.

When transformer trouble is indicated by missing ac voltage at the output, continuity and short tests of the transformer should be made. Using the ohms function of the EVM and a very low ohms scale, measure the resistance of the transformer windings. Although the exact correct resistance may not be known, a completely missing output will be caused only by a complete open or direct short. Shorts between the turns of one winding will reduce the output but not remove it. Using a high-resistance scale, also check the resistance between the windings and between each winding and the core.

A reduced dc output is usually a sign of a reduced input or of a heavy load being placed on the power supply. Pull fuse F1 and measure the ac voltage across L2. If this is normal, something is drawing too much current and loading the power supply. Leaky capacitors or low resistance in the load circuit are two common causes. Removing load resistance R_L by breaking the circuit between the top of C2 and the top of R_L can sometimes be helpful. Without the drain of R_L on the power supply, the dc voltage across C2 will be considerably higher than normal. If not, the capacitors are the source of the problem.

Care should be exercised when making resistance measurements in power supply circuits. The large capacitors used in the filter circuit are capable of retaining a large charge for some time after the power has been removed—especially in problems involving open load circuits. After removing power from the circuit, always short the terminals of the filter capacitors to ground with an insulated-handle screwdriver before making any resistance measurements.

TROUBLESHOOTING AMPLIFIERS

Successful troubleshooting of amplifiers using the EVM depends primarily on two factors. First, the meter must be accurate enough to measure the bias and signal voltages present in the circuit. Typical solid-state amplifiers (with silicon transistors) have about 0.6V across the base-to-emitter junction during conduction. Sometimes it is very difficult to measure this small voltage difference. Therefore, instead of measuring the voltage between the electrodes we measure between the base and ground, and the emitter and ground, as shown in Fig. 3-19. In the example shown, voltage measurements taken between the electrodes and ground show the emitter to be at —3.2V, while the base is at a —3.9V, a forward bias of 0.7V. This makes the base 2.4V more negative than the emitter. Under these conditions, if the actual difference in potential between base and emitter is measured, a similar determination can be made, although possibly with less accuracy. With the negative probe attached to the base

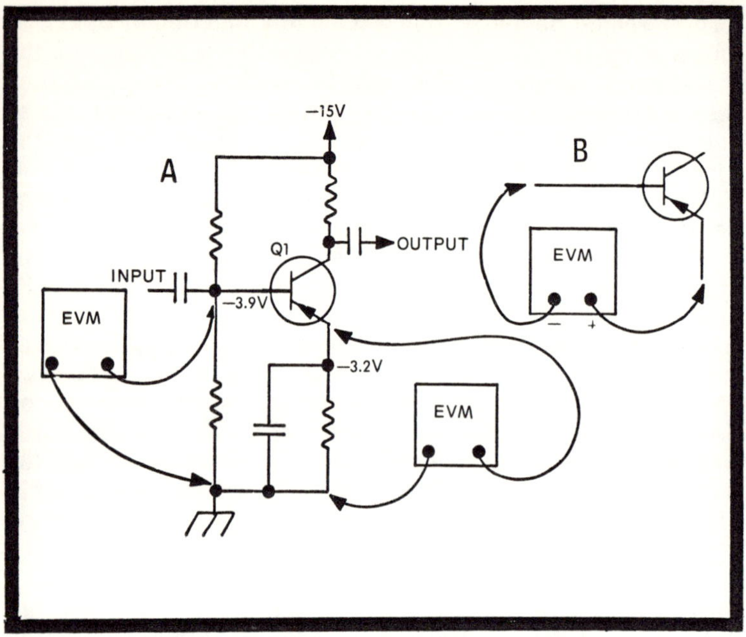

Fig. 3-19. Measuring transistor bias voltage.

and the positive probe attached to the emitter, the meter should read 0.7V, approximately.

The second most important troubleshooting factor, after meter accuracy, is the technician's knowledge of circuit operation. Knowing what to expect under varying conditions is an absolute necessity if accurate conclusions are to be made. More and more servicing diagrams are showing what voltages should be present where, but there are always cases in which proper circuit voltages must be "guesstimated." For this type of troubleshooting, there is no substitute for experience. To illustrate the techniques used with voltage analysis in amplifier circuits, we can refer to Fig. 3-20.

Amplifier Analysis

Amplification is performed by the pnp transistor, Q1. To isolate the amplifier from the dc voltages in the preceding stage, capacitor C1 is used to couple the ac signal to the base of Q1. Collector supply is furnished through R3, while R1 and R2 form a voltage divider for fixed base bias. Resistor R4 is the emitter stabilization resistor, and C2 bypasses the signal around the resistor to prevent signal degeneration. The action of R4 and C2 establishes a steady bias voltage at the collector. Capacitor C3 couples the signal from the amplifier to the next stage and, at the same time, isolates the dc voltage of the two stages.

The basic principle of voltage analysis of a transistor amplifier consists of noting the three key voltages (at the collector, emitter, and base) and determining the conduction level of the transistor. Then thought must be given to whether this is the normal operating condition of the circuit, or whether something is malfunctioning. Before we get to the actual analysis, let's look at some of the troubles that could occur and their effects on circuit voltages. Figure 3-20 is the schematic of an actual circuit I used for making the voltage measurements. An EVM was used to record voltages for both normal and faulty operation. Normal voltages are shown on the schematic.

Amplifier Defects

First look at the input coupling capacitor, C1. Under actual operating conditions, this capacitor couples the ac signal into the amplifier from the previous stage. In the test setup, a signal generator was used instead of a preamplifier. An open C1 was simulated by cutting one of the leads of C1. This removed all ac from the circuit; no change in dc voltages could be measured.

A leaky capacitor acts somewhat like a capacitor and a resistor in parallel. To simulate leakage in C1, the capacitor

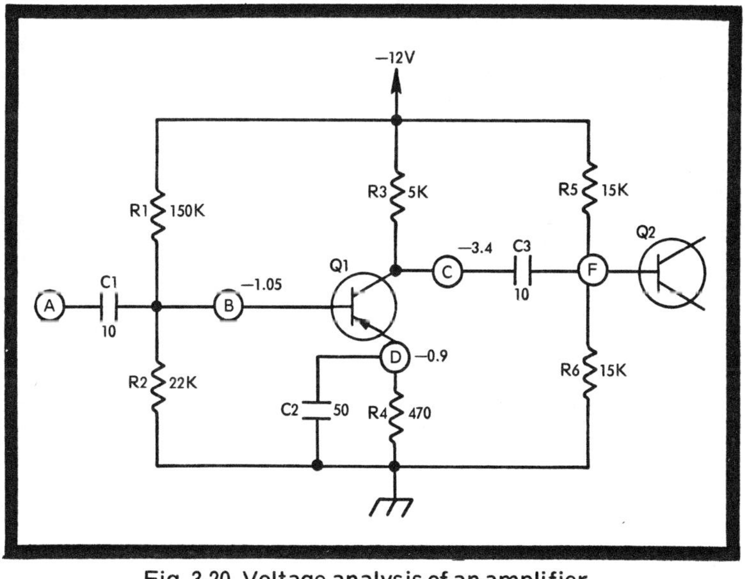

Fig. 3-20. Voltage analysis of an amplifier.

was shunted with various resistances. The dc of the preamplifier can be observed at the base of Q1 when leakage occurs. If the preamplifier has a low-impedance output with no output-coupling capacitor, the forward bias for Q1 will be reduced by the leakage, collector current will drop and collector voltage will increase. As the leakage of C1 approaches the point where the capacitor is effectively shorted, all output is lost. In the test setup, a signal generator with a high-impedance output with no dc present in the test signal voltage replaced the preamplifier. Under these conditions, with a 0.2V rms signal applied, the shunting resistance of C1 was decreased in discrete steps. With a 100K resistance shunting C1, there was a slight reduction in the output ac voltage at point F. The dc voltage at point B dropped to —0.91V, while the voltage at point D dropped to —0.8V, decreasing the forward bias by 0.04V. As a result, the output decreased slightly. As the leakage was increased, the base and emitter voltage continued to drop, and forward bias was decreased. Naturally, there was a corresponding decrease in the output voltage at point F. When the leakage resistance was increased to 10K, not only was there a reduction in the output voltage, but distortion of the sine wave was visible on an oscilloscope. As the leakage resistance approached 5K, distortion was quite obvious. When C1 was shorted, Q1 cut off, the collector voltage rose to 12V, and no ac output was present. However, a dc bias of 6V was present at F, since the base bias for the following stage was unaffected by this malfunction.

Perhaps a point should be clarified here about the output at point F. Under normal operating conditions, the output is a 1.5V rms sine wave when a 0.2V rms sine wave is applied at A. Care should be taken when interpreting the voltage at F. I measured the ac output here under normal conditions with an EVM that did not block the dc when set for the ac function, and got an 11.5V reading due to the effects of the 6V dc. When I made this same measurement with an EVM that isolated the dc, the reading was 1.5V ac. This illustrates the necessity of knowing your test equipment and how it responds under various test situations.

While we are discussing capacitors, let's go on to C2. With C2 open, ac degeneration occurs due to the increase in emitter voltage with an increase in signal input. This causes a severe reduction in the output.

A leakage situation similar to that used in connection with C1 was also established for C2. With a 100K leakage resistance simulated, there was little change in any circuit voltage or the output. As the leakage resistance reached 10K, positive clipping started to occur in the output, although there was only a slight change in circuit voltages. The greater the leakage of C2, the greater the positive clipping was.

The opening of C3 had no effect on the dc voltages within the amplifier of Fig. 3-20; however, the ac output at point F was missing. Leakage in C3 had more effect on the operation of the following stage than on Q1. A leakage in C3 had roughly the same influence on the following stage that a leaky C1 exerted on Q1. There was some reduction in the dc voltage at point C and the output voltage at point F.

Resistor Defects

Resistors R1 and R2 are the base bias resistors for Q1. If a malfunction upsets the resistance ratio of these two resistors, the bias and thus the conduction of Q1 changes. When the resistance of R1 is decreased, the base voltage increases, causing collector current to increase and collector voltage to decrease. The more R1 decreases in resistance, the greater the conduction level of Q1, and thus the lower the emitter voltage becomes. If R1 increases, a greater portion of the total voltage drop across the voltage divider is felt across R1. This causes a lowering of the base voltage, reducing the conduction of Q1, and collector voltage rises. The greater the resistance of R1, the greater the collector voltage. If R1 opens, conduction ceases and the collector voltage equals the supply voltage.

The resistance of R2 is also vital to the bias of Q1. As the resistance of R2 is decreased, the base voltage and conduction decrease. In some ways this is similar to the effect of a leaky C1. When R2 was completely shorted, Q1 was cut off and the collector voltage equaled the supply voltage. Increasing the resistance of R2 increased the voltage at point B. The greater the increase in R2, the more voltage at point B and the higher the conduction level of Q1; the output becomes more and more distorted. When R2 is open, conduction reaches maximum and the output is severely distorted. If R2 is allowed to remain open for any length of time, Q1 might be destroyed by the heavy current.

A collector load is furnished in this circuit by R3. Any change in the resistance of the collector load resistance will change the output. An increase in the value of R3 reduces the dc voltage applied to the collector of Q1, and a corresponding decrease in the output can be expected. The more the resistance of R3 is increased, the more the collector voltage will drop, until R3 is open and all collector voltage is removed. Now there is no dc path; therefore, Q1 does not conduct and no output is present at point F. A decrease in the resistance of R3 increases the collector voltage, causing an increase in conduction level. As a shorted condition of R3 is approached, the collector voltage and current reach a maximum; the signal output is low and distorted. Under these heavy-current conditions, Q1 may be destroyed.

The forward bias of Q1 will be altered with any change in the resistance of R4. When R4 was shorted, the effects were the same as a short across C2. In this particular circuit, a shorted R4 causes an increase in the output, collector current increasing and collector voltage decreasing. When R4 opens, the dc path is removed, no collector current flows, and the collector voltage reaches the level of the power supply.

It is relatively easy for a technician to look at a schematic of a circuit and describe the probable symptoms that will result from various component failures. Unfortunately, we cannot pick our problems in this manner when it comes to actual troubleshooting. We must work from the symptoms and try to locate the cause. A symptom such as loss of output is not specific enough to pinpoint the source of trouble. More information is necessary. This information can be gained by voltage checks in the amplifier circuit. Three key dc voltages for this analysis are the emitter, collector, and base voltages of the transistor itself. Start with the collector voltage.

Amplifier Voltage Checks

In the example in Fig. 3-20, the normal collector voltage was —3.4V. An increase in that voltage would indicate a decrease in the conduction of Q1. Conversely, a decrease in voltage at point C would indicate an increase in conduction. Looking back over the possible failures we have already discussed, we see that an increase in collector voltage could be caused by several different component failures: a decrease in the forward bias voltage at point B, a decrease in the value of R3, an increase in the negative voltage at point D, or a malfunction in C1 itself causing a decrease in condition. My first response in this case would be to measure the base voltage. If too little forward bias is causing the collector voltage to rise, the voltage at B will be less negative than during normal operation. Should low base voltage be the problem, then we are down to one of three likely possibilities: a leaky C1 draining off the bias voltage through the preamplifier stage, a decrease in the resistance of R2, or an increase in the value of R1. With a low-power EVM such as the Sencore FE20, the resistance of R1 and R2 may be checked in the circuit, and a similar measurement may be made of the leakage of C1. An alternative to checking the leakage of C1 is to lift one lead and measure the voltage at point B under an open-capacitor condition. If the base voltage returns to normal, the capacitor was leaky. Also, the ac voltage across C1 can sometimes be used to determine if the capacitor is doing its job; under normal conditions this should be a very small voltage. A decrease in the value of R3 would also increase the collector voltage; however, the base bias voltage would be normal and the voltage at point D would increase due to the added current through the transistor. An increase in the emitter voltage could also be the result of an increase in the value of R4.

An open transistor can be the cause of high collector and low emitter voltage; base voltage will be close to normal. This will very nearly duplicate the symptoms of an open R4. To determine if the transistor is actually at fault, connect an EVM to read collector voltage; then jumper a resistor of about 100K, momentarily, between the base and collector. This will place a definite forward bias on the base, and the transistor should conduct, causing collector voltage to drop. If not, the transistor is the cause of the high voltage at point C.

Just as an increase in collector voltage was an indication of too little conduction, a decrease in collector voltage is a good indication of too much conduction through the transistor. Again, this might result from several different component failures. If base bias voltage measurements indicate too high a base voltage, look for trouble in the base bias voltage divider. A short across R4 and C2 could also be the cause. Shorting of Q1 would also cause a drop in the collector voltage. To determine if the transistor is at fault, connect an EVM to measure collector voltage. With the low-voltage indication present at point C, momentarily short the emitter to the base with a test lead. If the meter shows an increase in voltage, a bias problem exists in either the emitter or base circuit, and the transistor is probably okay.

In some troubleshooting situations, voltage measurements will not reveal the source of trouble in an amplifier circuit, and other tests using an oscilloscope and a signal generator may be necessary. However, the EVM can be used to locate most any trouble in a solid-state amplifier in which the circuit dc voltages are upset. For this reason, the EVM can be used to great advantage in troubleshooting solid-state amplifiers.

TROUBLESHOOTING OSCILLATORS

Many of the same techniques of voltage analysis that are used in troubleshooting amplifiers may be used to locate faulty components in oscillators. In order for oscillation to occur in the circuit shown in Fig. 3-21, the voltage divider

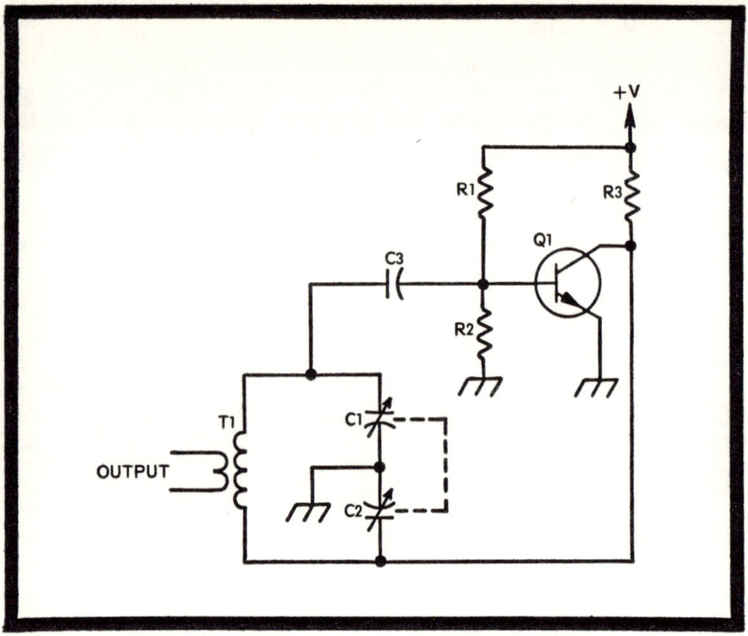

Fig. 3-21. Voltage analysis of an oscillator.

composed of R1 and R2 must furnish forward bias to transistor Q1. With Q1 conducting, energy is applied to the tank, and feedback through C3 to the base causes oscillation. Capacitor C3 also blocks the flow of direct current through the tank circuit. Resistor R3 furnishes the necessary collector load. The output waveform is taken from the secondary of T1.

The primary of T1, C1, and C2 control the frequency of operation. Any change in value of these components will change the frequency of the output. Partial shorts—that is, shorts between individual turns of the secondary of T1—can cause a decrease in the amplitude of oscillations and a shift in frequency. A complete short across the transformer or the tuning capacitors will cause oscillations to cease. A decrease in the conduction of Q1 will be evidenced by an increase in the collector dc potential. The bias voltage at the base will also change. An open in these components will cause similar symptoms.

Capacitor C3 performs the same function as the input coupling capacitor in the amplifier circuit of Fig. 3-20. Malfunctions will have the same general effect on the bias of the transistor. An open C1 will remove the feedback signal from the tank, and oscillation will cease. Just how a leaky C3 will change circuit operation will depend upon the amount of leakage present. Any leakage will cause some of the bias current to be drawn off through the tank circuit. This can cause a shift in frequency, a decrease in the output, or a complete loss of output, depending on the seriousness of the leak.

Voltage divider R1R2 is very critical to the entire operation of the oscillator. Should R1 open, forward bias is removed and the oscillator will not start. However, if oscillations are started by some external means, or R1 opens after oscillations have begun, the circuit will continue to operate. A shift in frequency or a decrease in amplitude may be noted. A decrease in the resistance of R1 lowers the base bias. When R1 is completely open, the collector current drops, the collector voltage rises, and the output is lost. If the collector current is only slightly low, the amplitude of the output is down a little, and the oscillator is unstable, look for an increase in the value of R2. A decrease in the value of R2 causes collector current to decrease. If R2 becomes shorted, the collector current drops to zero and there is no output.

The collector load resistor of the oscillator has a great similarity with that of the amplifier. As R3 increases in value, collector current decreases; when a complete open is achieved, no collector current can flow and oscillations cease. If R3 decreases in value, the signal amplitude is decreased. A direct short across R3 causes a complete loss of feedback and, therefore, a loss of output.

Problems with Q1, of course, will affect the operation of the oscillator. In-circuit methods of testing the transistor by changing the base bias voltages may be used with oscillators as with amplifiers.

4 Fundamentals Of Waveform Measurements

A waveform is a pictorial representation of a varying potential plotted against time. By using a linear time base, an unknown waveform may be plotted and analyzed. This analysis is one of the most important steps in troubleshooting, since it supplies much information about the characteristics of many electronic circuits. The path of a signal and its waveshape may be traced through a circuit and the gain and distortion characteristics of the circuit thereby determined. Or if the circuit is defective, the particular stage causing the trouble may be isolated. The waveform of the signal can indicate the presence of harmonics or oscillations, or it can indicate just how well the circuit is doing the job it was designed to perform. From this we can see that the oscilloscope is one of the most important pieces of test equipment.

An oscilloscope incorporating only the basic circuits necessary for operation has been selected for discussion in the following paragraphs, in order to facilitate explanation. Early paragraphs will provide information on a few circuits required to make a useful, although limited, test instrument. As the need for more accurate and special oscilloscope presentations arises, various circuit refinements and modifications are added to the basic oscilloscope. Today's applications, especially television circuit analysis, require that the oscilloscope bandpass be very broad so as to faithfully reproduce the signals being observed. This resulted in the development of the modern wideband triggered-sweep oscilloscope, which will be discussed later.

THE BASIC OSCILLOSCOPE

The cathode-ray oscilloscope provides a visual representation of some electrical quantity as a function of another quantity on the screen of a cathode-ray tube (crt). The usefulness of the oscilloscope is inherent in its ability to graphically portray instantaneous fluctuations in circuit waveforms. The operation of the oscilloscope is based on the control of a beam of electrons used to draw the shape of the waveform on a fluorescent screen. Since the electron beam itself is used to inscribe the outline of the waveform, the crt responds to much higher frequencies than other electrical indicating devices. The major components of a basic oscilloscope include a crt, a sweep generator circuit, deflection amplifiers, and the necessary controls and switches for proper operation.

The CRT

The very heart of the oscilloscope is the crt. A simple form of the crt is illustrated in Fig. 4-1. Electrons are emitted by the heated cathode and accelerated by the attraction of the second anode. In early tubes, there was little control of either the number or the direction of the electrons emitted from the cathode. Modern tubes have overcome the disadvantages of earlier types by adding control grids, focusing grids, and other elements.

Electrostatic control means that deflection and focusing in the crt are accomplished by electrostatic fields that exist between the electron beam and the elements of the crt. Electrons emitted by the cathode are focused (Fig. 4-1) and accelerated by the voltages present at the various grids and anodes. The apertures in the tube elements and the static fields around the anodes form the electron stream into a narrow beam, which passes between each of the two sets of deflection plates before reaching the screen. The beam is moved about the screen by varying the potentials applied to the deflection plates. Varying the potential between the vertical plates moves the beam up and down the screen, while the voltage at the horizontal plates controls the back-and-forth motion of the beam.

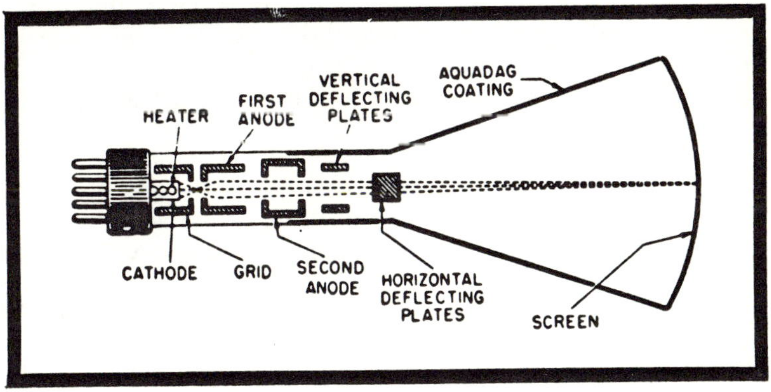

Fig. 4-1. A basic crt.

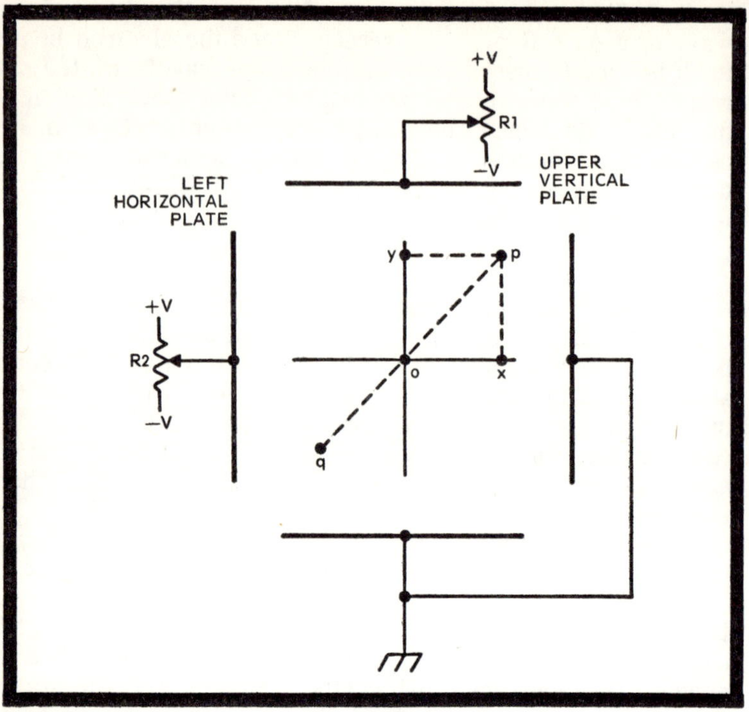

Fig. 4-2. Electron beam motion.

With no potential applied to either pair of deflection plates, the beam is not deflected and remains stationary, producing a spot of light in the center of the screen. When an alternating signal is present between the vertical plates, the beam is deflected first toward one vertical plate and then toward the other. Since there is no deflection in the horizontal direction, a vertical line is formed at the center of the screen. Removing the signal from the vertical plates and connecting it to the horizontal plates will result in a horizontal line across the screen. When signals are applied to both pairs of deflection plates at the same time, the deflection of the beam is the resultant of the vertical and horizontal forces at any instant. For example, if both the top vertical plate and the right horizontal plate have the same positive potential at the same instant, the spot will be deflected toward the upper right corner of the screen.

This action can be clarified by referring to Fig. 4-2. The deflection plates act upon the beam independently and simultaneously, producing motion of the spot along a vector that is the resultant of the forces exerted on the beam by the potentials of the plates. In this manner, the electron beam is continually acted upon by two forces which are at right angles to each other. The figure illustrates the resultant motion produced by the independent deflection voltages applied at the same time to the **X** and **Y** axes.

When the wipers of potentiometers R1 and R2 are at ground potential, all four plates are grounded and the spot appears at point **o** on the screen. If the wiper of R2 is moved toward the negative voltage, the horizontal plate on the left becomes negative with respect to the one on the right. This causes the beam to be repelled by the left horizontal plate, the spot moving to point **x** on the screen. Now, if we move the slider of R1 toward the positive terminal, the upper vertical deflection plate becomes positive with respect to the lower plate, and the beam is attracted from point **x** to point **p**. If you were quick with your hands and had the coordination of a neurosurgeon, you could move R1 and R2 simultaneously at the same rate, causing the spot to move along line **op**. Or the controls could be varied to make the beam move along line **oq**.

From this demonstration you can see that two voltages—steady or variable—applied simultaneously, one to each pair of deflection plates, will cause the spot to be positioned at any instant according to the resultant of the two forces.

One of the most useful applications of the oscilloscope is for the observation of signal variations with respect to time. The most common method of representing voltage waveforms uses the vertical axis to represent the amplitude of the voltage and the horizontal axis to represent time. For this, a voltage must be impressed on the horizontal deflection plates which will move the electron beam from left to right at a constant rate of speed. The beam strikes the screen at only one point at any instant, but it is possible to form a line by rapidly producing a succession of many spots of light close together. The human eye retains any one image for a very short time, so a motion which is fast enough appears as a blur, because successive images overlap. If the horizontal deflection voltage causes the spot to move fast enough, the image will be a line which appears stationary on the screen. The voltage that causes this horizontal deflection is called a **sweep** voltage, because it sweeps the spot across the screen. The line which results is called the **time base**, since a given distance along this line will represent a specific period of time.

The time base generator, or sweep oscillator (Fig. 4-3), produces a waveform similar to that in Fig. 4-4. This **sawtooth** waveform is applied between the horizontal deflection plates. At point A, the spot is at the left side of the screen. Between

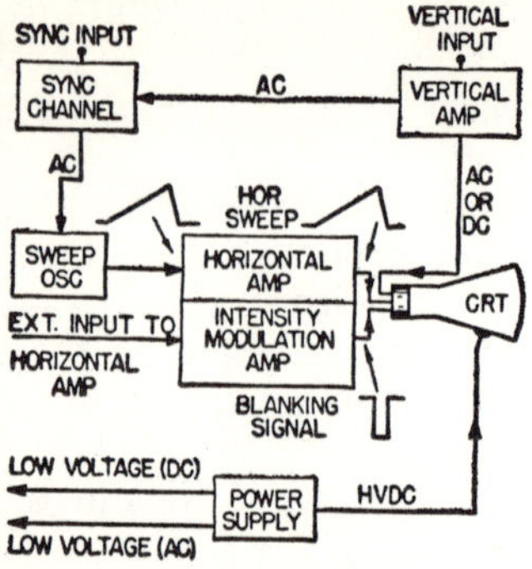

Fig. 4-3. Basic oscilloscope.

points A and B, the spot is moved in a linear manner across the screen from left to right. At point B, the spot is at the left side of the screen. During time B to C of the waveform, the spot is returned rapidly (note the steepness of the waveform) to the right side of the screen so that it is ready to start the next sweep. On some scopes, the beam is turned off during this retrace time to prevent the retrace line from being confused with the actual time base line.

During the time the sawtooth waveform is being applied to the horizontal plates, if a sinusoidal signal is applied to the vertical plates the trace on the screen of the crt will be the outline of a sine wave, as shown in Fig. 4-5.

At zero degrees, the sawtooth, which is applied to the horizontal plates, has moved the beam to the left side of the screen. At this time, the sine wave is at the zero-amplitude point. As the sawtooth waveform starts its sweep across the tube (from left to right), the sine wave begins to deflect the beam toward the top of the screen. At 90 degrees, the sawtooth has completed one-fourth of its sweep, and the sine wave has deflected the beam to its maximum direction extent.

When both deflection voltages reach 180 degrees, they are passing through their zero reference lines. With equal potentials on all plates, the spot will be in the center of the screen. As the sawtooth waveform continues to move the

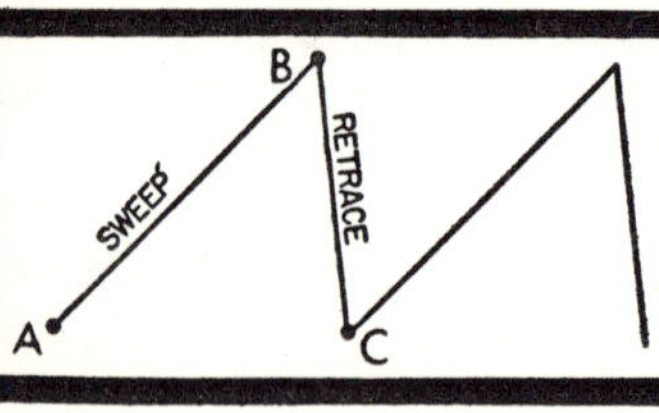

Fig. 4-4. Sawtooth sweep voltage.

beam toward the 360-degree position on the screen (far right), the sine wave deflects the beam toward the bottom of the screen. At 360 degrees, the sawtooth sweep retraces the beam rapidly back to the left side of the screen.

Vertical and Horizontal Amplifiers

The deflection plates of a crt are relatively insensitive; that is, they require a fairly large signal to deflect the beam. As shown in Fig. 4-3, it ordinarily is necessary to amplify both the vertical and horizontal signals.

When an amplifier is used between the signal source and the crt plates, the quality of signal reproduction is dependent upon how faithfully the amplifier reproduces the input signal. Limitations of fidelity include frequency discrimination, phase distortion, and maximum input voltages. One very important consideration in choosing an oscilloscope is the frequency response of the input amplifiers. Many applications of the oscilloscope call for the observation of pulse and square-wave voltages. Therefore, not only must sine-wave response be uniform, but the **transistor response** must be broad enough to permit undistorted amplification of these waveforms.

The Sweep Generator

When observing any waveform, keep in mind that an unknown quantity is always plotted as a function of a known quantity. In most cases, the horizontal axis is the known quantity—time. When the rate, or frequency, of the sweeping horizontal sawtooth is adjusted to synchronism with the

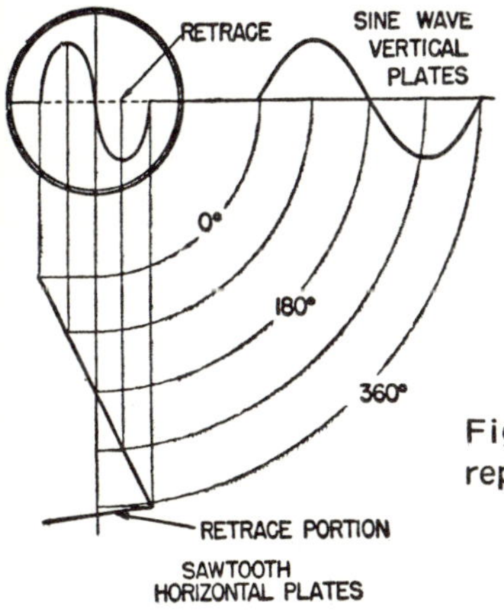

Fig. 4-5. How a pattern is reproduced on the crt.

unknown signal applied to the vertical plates, the time variation of the voltage is traced on the screen of the crt. The sawtooth sweep voltage is produced by the sweep oscillator circuits, as shown in Fig. 4-3.

The shape of the sawtooth sweep voltage is based on charging a capacitor through a resistance from a constant potential to obtain an exponential charging wave that is linear enough for time base application. After charging to a predetermined level, the capacitor is quickly discharged, and the entire process is repeated at a frequency determined by the RC time constant of the circuit.

Sweep voltages obtained in this manner do not produce a sweep of sufficient linearity for some applications, because the charging voltage follows an exponential curve instead of a straight line. By replacing the RC network with a circuit having a more linear output, better sweep linearity can be obtained.

A stationary display may be obtained by adjusting the time base generator to some harmonic (or subharmonic) of the unknown signal applied to the vertical amplifier. Due to fluctuations in either the sweep or test signal, the two signals often lose synchronism. This causes the waveform to move on the face of the crt, which can make the trace unintelligible. One method of locking the sweep generator to the frequency of the test signal is to inject a portion of the test signal into the sweep circuits. This is called synchronization.

OSCILLOSCOPE OPERATION

While we are still on the subject, let's discuss some of the operating controls and features found on a basic oscilloscope. These controls can be divided into three major groups according to their effect on the oscilloscope operation, as shown in the front panel diagram in Fig. 4-6.

Beam Controls

There are usually several adjustments included for beam control: **vertical position, horizontal position, intensity,** and **focus.** The **position** controls adjust the pattern on the crt screen vertically and horizontally for locating the trace in the most advantageous position for viewing.

While the **position** controls affect the overall display of the waveform, the **intensity** and **focus** controls regulate the characteristics of the electron beam itself. The **intensity** control regulates the number of electrons in the stream and thereby controls the brightness of the trace. The **focus** control adjusts the diameter of the beam, allowing for a sharply defined trace.

Vertical Controls

The vertical gain control regulates how much the vertical amplifier boosts the input signal. On more advanced oscilloscopes, you will find a more sophisticated vertical control system.

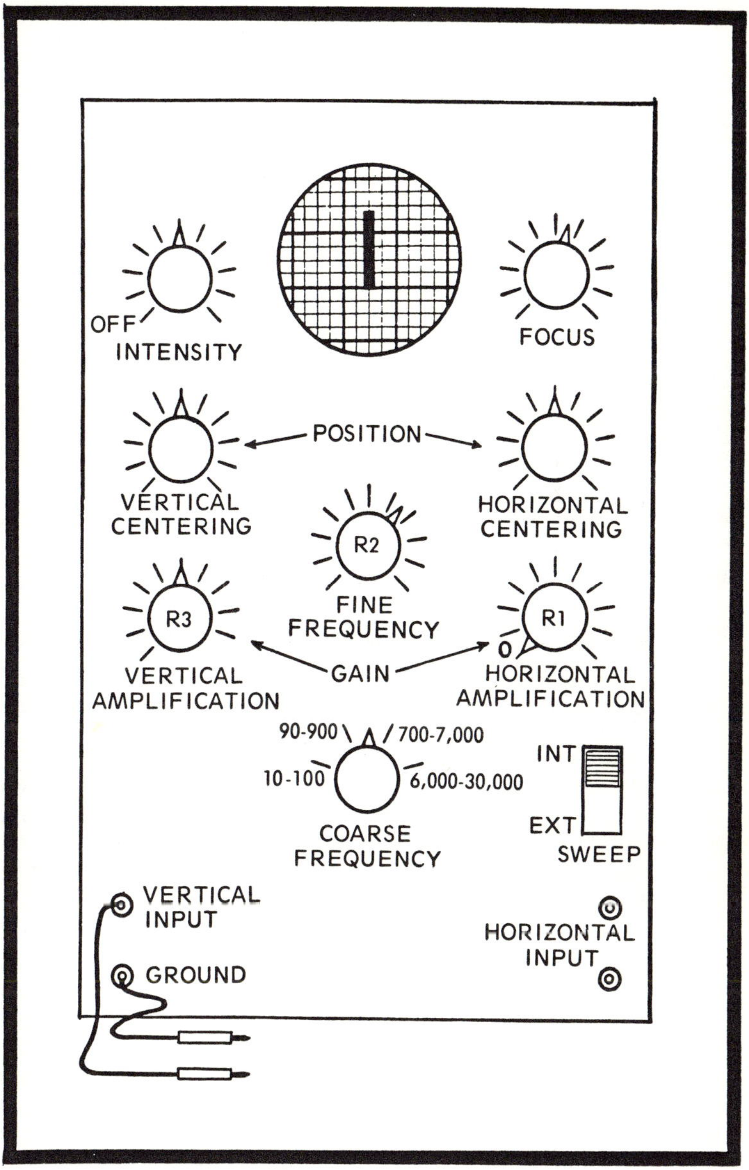

Fig. 4-6. A simple oscilloscope front panel.

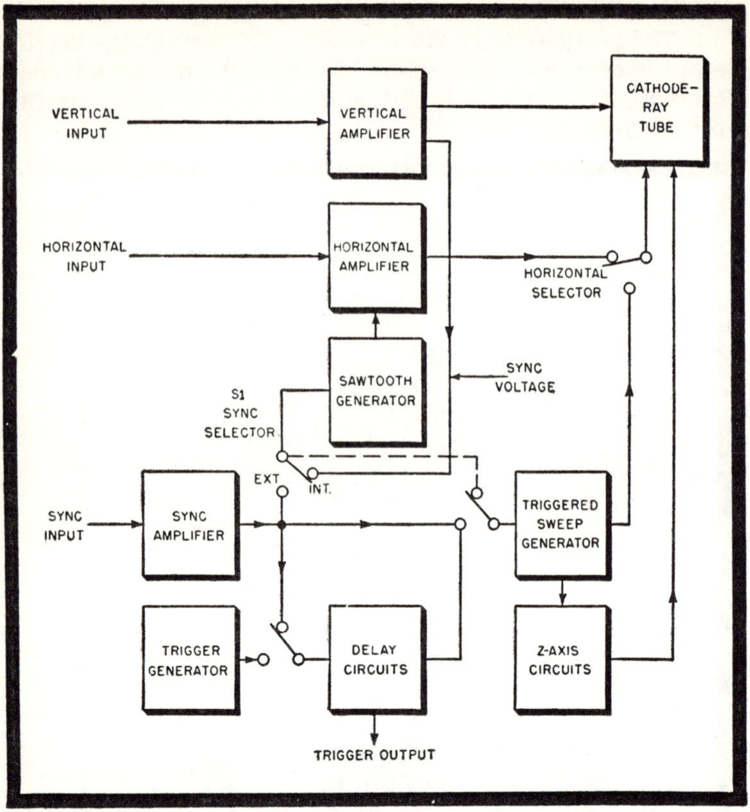

Fig. 4-7. Simplified block diagram of an oscilloscope.

Horizontal Sync and Sweep Controls

This is the most difficult group of controls for the beginner to understand. (Refer to Fig. 4-7.) The simplest of these controls is the **horizontal gain**, which regulates the amplification of the horizontal amplifier and the speed of the horizontal sweep on the crt. The **gain** control is represented by R1 in Fig. 4-6.

Most oscilloscopes provide for the horizontal sweep voltage to be either generated internally by the sweep generator or supplied from some external source. That is the function of S1 in Fig. 4-7 (**horizontal sweep** in Fig. 4-6). When S1 is set to the **external** position, the internal sweep circuits are disconnected from the horizontal amplifier, and the amplifier receives its sweep signal from the **horizontal input-external sync** jack. In this mode, the oscilloscope plots the amplitude of the vertical signal against a time base generated by an external sweep system. The crt deflection plates are effectively connected to two separate signals through the vertical and horizontal amplifiers, as illustrated in Fig. 4-8. In this mode, two external signals may be compared, or the vertical signal may be plotted against a time base generated by an external device.

SINUSOIDAL SWEEP

Assume that two separate sine waves are applied to the horizontal and vertical input terminals of the oscilloscope (Fig. 4-9). If both inputs were at their zero reference line at the same instant, the crt beam would be positioned at the center of the screen. In this discussion, though, let's say that a 90-degree phase difference exists. As the vertical signal is passing the zero reference, traveling from negative to positive (point 1), the horizontal signal is reaching its maximum positive amplitude (point 1). Here we must assume that the signals are of equal amplitude and the gain of the horizontal and vertical amplifiers are identical.

With these conditions established at time 1, the electron beam will be positioned to the far left along the horizontal axis. The vertical signal is at zero, so no vertical deflection occurs; however, there is maximum horizontal voltage present, and the beam is pulled over to the indicated position. The horizontal signal now starts to decrease, and the vertical signal starts to increase in the positive direction. Since the horizontal signal is decreasing, the beam starts to move toward the center of the screen. At the same time, the vertical

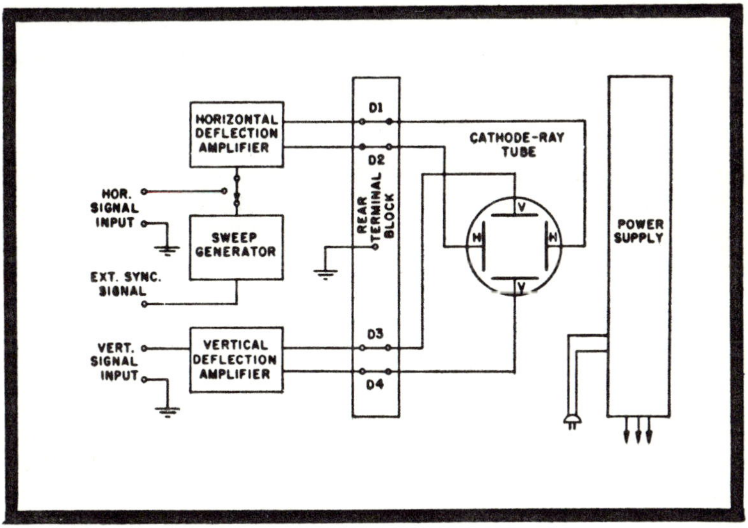

Fig. 4-8. External horizontal input mode.

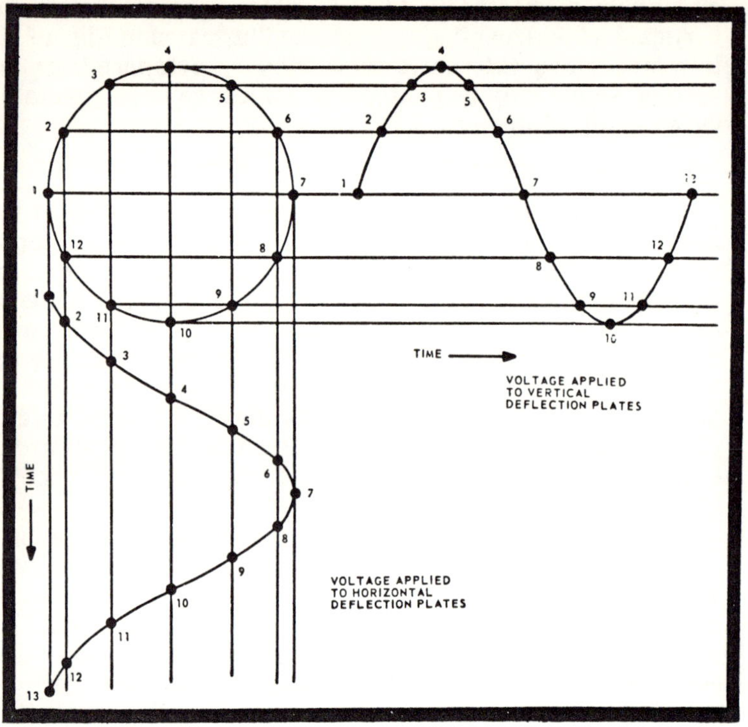

Fig. 4-9. Oscilloscope display using an external sweep.

deflection signal is increasing and pulling the beam upward. The combined effect of the changing signal voltages pulls the beam toward the top and to the right (times 2 and 3).

The vertical signal continues to increase and the horizontal signal continues to decrease. At time 4, the vertical signal is at its maximum positive value, while the horizontal signal is passing through zero. These combined voltages at the deflection plates position the electron beam at point 4 on the circle.

At this time, the horizontal signal begins its negative half-cycle, while the vertical signal starts to decline from its maximum positive value. Combining the voltages at the crt deflection plates results in the electron beam being pulled high to the right, to point 5.

When the time advances to time 7, the vertical signal is passing through its zero reference line and the horizontal signal reaches the maximum negative amplitude. With no vertical voltage present at this instant, the electron beam is positioned along the horizontal axis at the maximum negative position (far right). As the vertical signal now begins its negative half-cycle, the horizontal signal is starting to return to zero after its maximum negative excursion. This pulls the beam downward and toward the vertical axis.

At time 10, the vertical input signal has reached the maximum negative value and the horizontal signal is going through zero, passing from the negative half-cycle to the positive half-cycle. This positions the electron beam at point 10 on the screen, along the vertical axis.

Time 12 sees the waveforms almost returned to the starting point. Although we have looked at discrete beam positions here, the beam actually is moving in a sweeping circular motion characteristic of sine-wave signals of the same frequency and 90 degrees out of phase. Due to the persistence of the crt screen and of vision, a circle appears on the screen. Patterns produced by applying sine-wave signals to both the horizontal and vertical inputs are called **Lissajous** figures.

Lissajous Figures

The simplest Lissajous figure is a circle. As we have seen, this is produced by two signals having the same frequency and amplitude, but differing in phase by 90 degrees.

Lissajous figures produced when two different frequencies are applied to the vertical and horizontal deflection plates are illustrated in Fig. 4-10. These patterns result when the vertical frequency is twice the horizontal frequency. It doesn't make any difference just what the vertical and horizontal frequencies are. If their **ratio** is 2:1, the bowtie figures of Fig. 4-10 will result. Notice how the phase difference between the vertical and horizontal signals changes the Lissajous figures obtained with a 2:1 ratio.

Two voltages in phase result in the neat bowtie shown in Fig. 4-10A. When the phase angle is 90 degrees, the loops are closed, as shown in C. This figure is a parabola. For phase angles greater than 180 degrees, the display is inverted, as shown at D.

Determining Frequency Ratios

We could draw a great number of pictures for different frequency ratios, but there is an infinite number of possible ratios and pictures. However, by examining the illustration in Fig. 4-10, it is possible to deduce a general rule by which we can determine any frequency ratio from the pattern it generates. In this regard, notice in Fig. 4-10 that the figures in parts A, B, and D have two lobes in the horizontal plane and

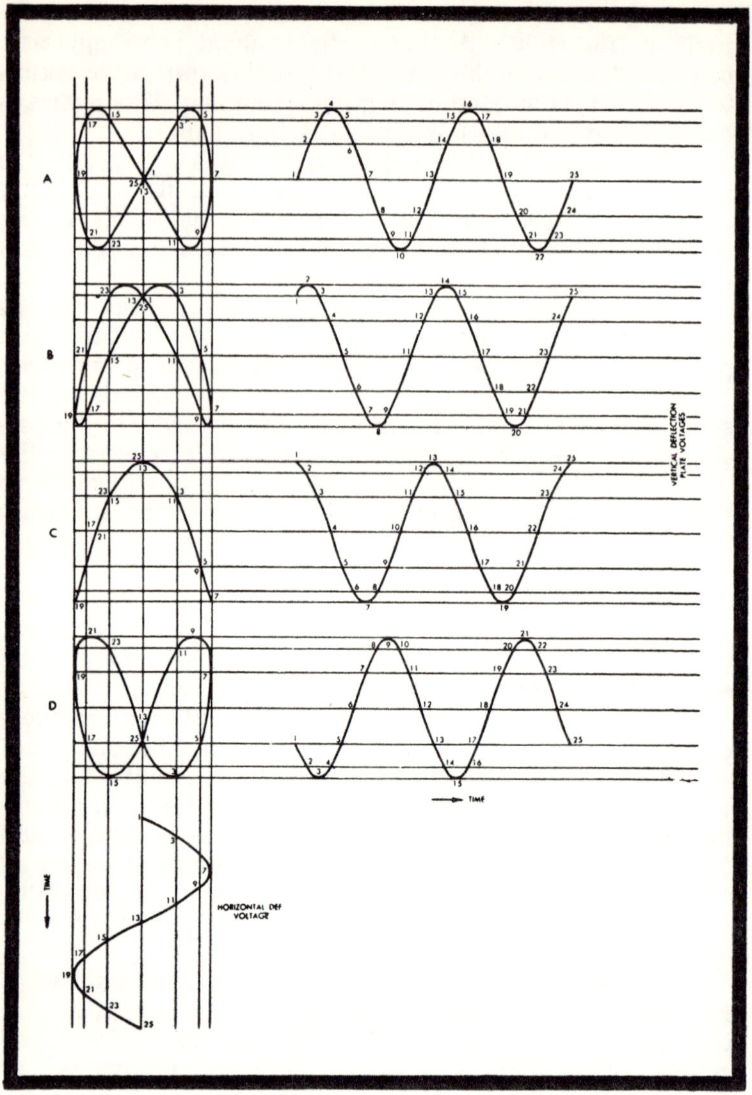

Fig. 4-10. Oscilloscope display with a frequency ratio of 2:1.

one lobe in the vertical plane. From this observation, you can form the following general rule: The ratio of the number of points at which a Lissajous figure touches a horizontal line to the number of points at which it touches a vertical line is equal to the ratio of the vertical frequency to the horizontal frequency. And the horizontal line can be at the top or bottom of the figure; the vertical line at the left or right.

In applying the rule to the pattern at C in Fig. 4-10, note that the luminous spot passes through point 25 (or, same thing, point 1) twice in tracing out the pattern. Thus, although the pattern appears to touch the horizontal line only once, it should be considered to actually touch it twice.

Sync Controls

The oscilloscope is most frequently used to display an external signal received via the **vertical input** terminals. This is plotted against an internally generated time base. To use this method of waveform observation, S1 and S2 in Fig. 4-7 are placed in the **internal** position. Functionally, the scope then is as shown in Fig. 4-11. The waveform to be displayed is routed from the vertical input terminals through the vertical amplifiers and to the vertical deflection plates. The horizontal sweep circuit generates a sweep signal that is shaped into a sawtooth before it is applied to the horizontal deflection plates. In this mode of operation, the sweep circuit operates at a frequency determined by the setting of the **frequency adjust**

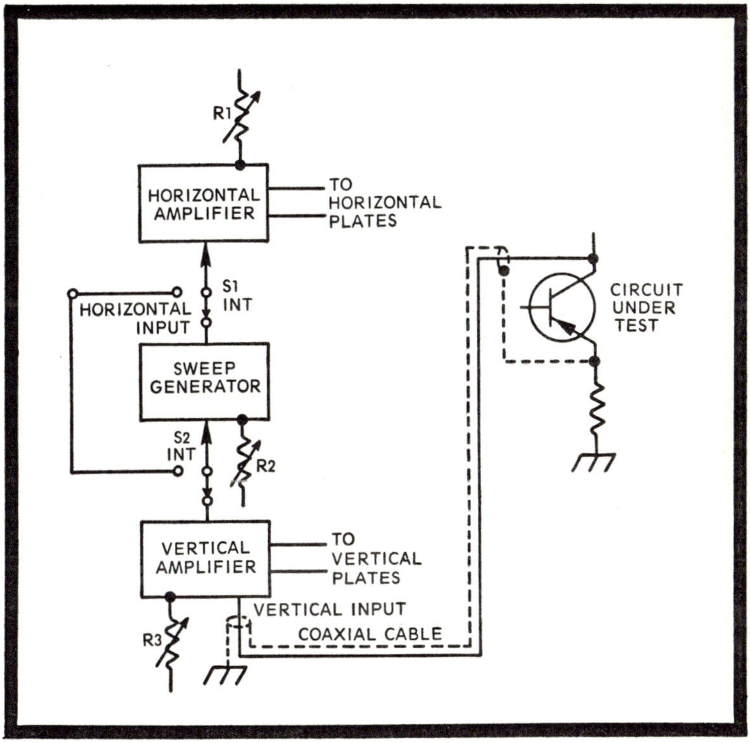

Fig. 4-11. Internal sync mode.

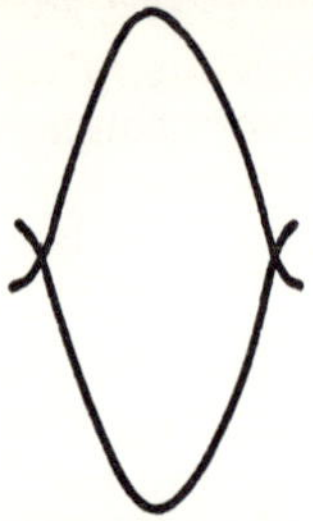

Fig. 4-12. Out-of-sync display.

control, R2. A portion of the vertical input signal is used to synchronize the sweep circuits. This is called **internal synchronization.** On most scopes, an adjustment is also provided in the sync circuits to control at what amplitude of the incoming signal the sync pulse is taken.

Using internal sweep and sync is very simple. Switches S1 and S2 are both placed in the **internal** position and power is applied to the scope. The sweep generator starts to "**run free,**" oscillating at its natural frequency as determined by the setting of **frequency adjust** control R2. This produces a straight horizontal trace across the crt. The positioning controls are used to set the trace at the desired position, and the beam controls are used to adjust brightness and focusing. The horizontal gain is adjusted to give the trace a convenient length. With this setup we are now ready to display a waveform.

Assume that R2 is at such a position that the sweep generator is running at approximately twice the frequency of the signal applied to the vertical input. Under these conditions, it might be hard to obtain good sync. The waveform will probably appear to be moving, and will look something like that shown in Fig. 4-12. What happens is that the signal applied to the horizontal plates is deflecting the beam across the crt screen twice during each complete cycle of the input signal. Under these conditions, half of the input wave is drawn during

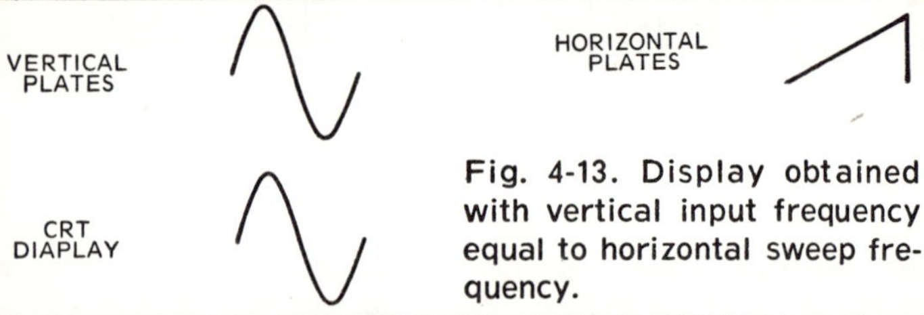

Fig. 4-13. Display obtained with vertical input frequency equal to horizontal sweep frequency.

each horizontal sweep. This causes the two half-cycles of the input waveform to be superimposed. Usually, under these circumstances, it is very hard to obtain a synchronized display; the waveform is very jittery and appears to move across the screen.

When the frequency control is set so that the sweep circuits are running at the same frequency as the vertical input signal, the display shown in Fig. 4-13 results. For each complete cycle of the vertical input signal, the horizontal sweep makes one complete cycle and one complete cycle of the input waveform is displayed. If the oscilloscope is operating normally, this type of display is not difficult to achieve.

Figure 4-14 goes one step further and illustrates the results of the sweep circuits being set at one-half the vertical input signal frequency. Now the beam is moved across the screen only once during two complete cycles of the input voltage. A display showing two complete cycles of the input waveform results. This is a typical display obtained when operating a service-type oscilloscope.

Other displays can also be used, showing **several** cycles of the input waveform. In each case, the sweep circuit is set to a subharmonic of the input signal. Once a display has been achieved, the vertical and horizontal gain controls, as well as the frequency and positioning controls, can be touched up to give the most meaningful display.

The final mode of operation to be considered for the basic oscilloscope is **internal** sweep and **external** sync. As mentioned previously, the internal sweep and sync mode sometimes presents an unstable display when observing complex waveforms. When this occurs, it is sometimes helpful to use a sync pulse from an external source. In this mode, the vertical input signal is plotted against a time base established by the oscilloscope's sweep circuits, and the synchronization of the sweep circuits is now controlled by some external source other than the input signal.

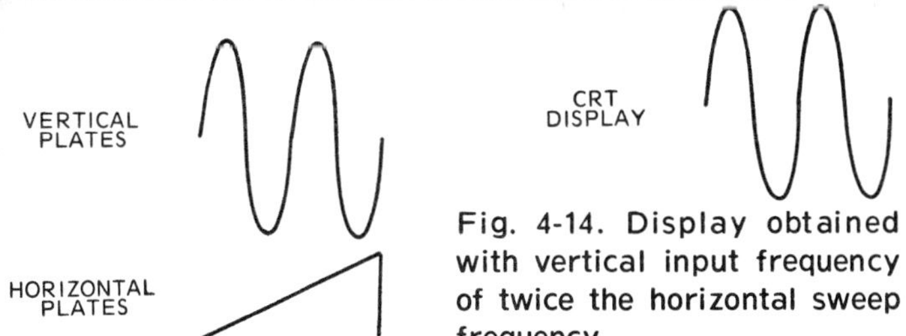

Fig. 4-14. Display obtained with vertical input frequency of twice the horizontal sweep frequency.

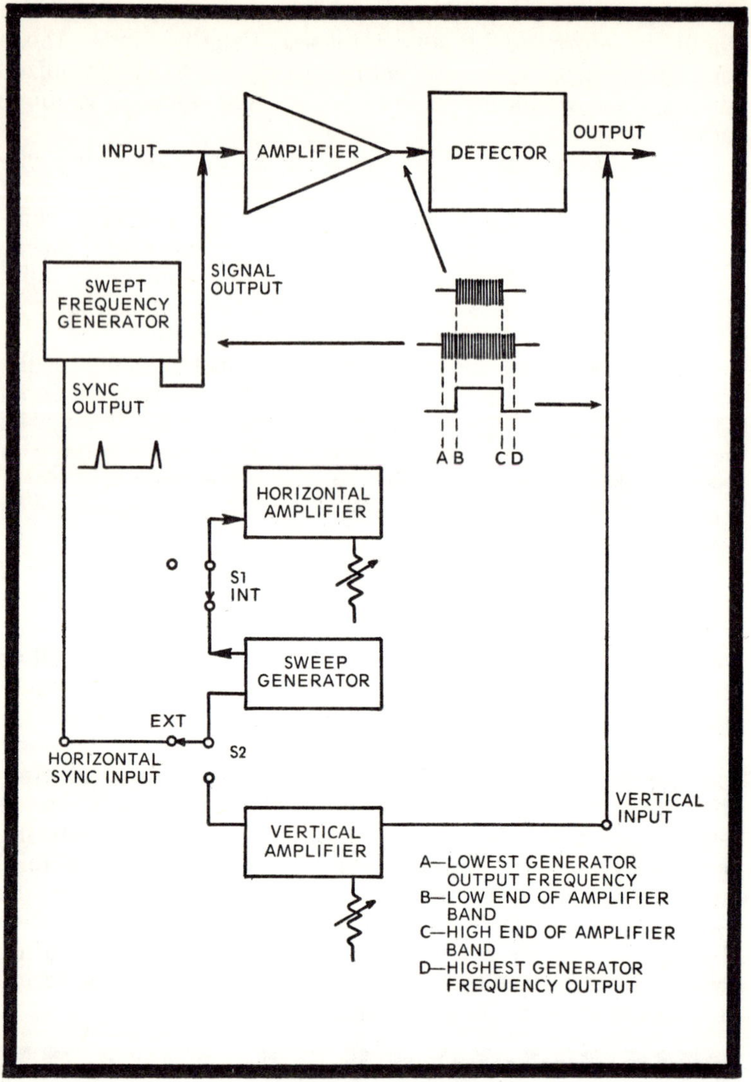

Fig. 4-15. External sync mode.

In Fig. 4-15, a **swept-frequency generator** is being used in conjunction with a basic scope to check the bandwidth of an amplifier. This generator is nothing more than a conventional signal generator whose output frequency is continually varied by an internal sweep generator (similar to the one in the scope). The output of the generator is then swept across a band of frequencies. Each time the signal generator starts its sweep across the band, its sweep circuit puts out a sync pulse.

This sync pulse can be used to control the scope's sweep generator.

With a setup as shown in Fig. 4-15, assume that the swept-frequency generator starts to sweep across a band of frequencies, starting below the cutoff frequency of the amplifier and continuing to a frequency above the amplifier's high-frequency limit. If the amplifier's response is linear over its designed frequency band, the output is as shown. The detector removes the high-frequency components of the signal as part of its function. This leaves only the bandpass envelope of the amplifier's output, which is applied to the scope as a vertical input.

At the time the frequency generator starts a sweep of the amplifier frequencies, a sync pulse is sent to the scope. The scope's sweep generator is thereby synchronized in such a manner that the time base of the scope is at the same time base as the sweep of the frequency generator, giving a stable display of the amplifier's frequency response.

SWEEP SYNCHRONIZATION

The previous discussion considered three popular methods of oscilloscope sweep control. The method by which the sweep circuit can be synchronized to the signal to be displayed is a very important factor when selecting a scope for a particular use. For example, external **sweep** is much more limited in application than external **sync**. And, although more versatile than the external **sweep** mode, the **external sync** mode is more limited than the **internal** sync mode. In some applications, a fourth method, called **triggered sweep,** is desirable.

Repetitive Sweep

Repetitive sweep is based on a synchronized free-running multivibrator. A basic free-running multivibrator circuit is shown in Fig. 4-16. The circuit is a simple two-stage resistance-coupled amplifier in which the output of the second stage is fed back to the input of the first stage. Since inversion occurs in each stage, a signal change at the grid of either tube will be amplified, inverted, and coupled back ot the grid of the other tube, where it is again amplified, inverted, and sent back in phase to the grid at which it originated.

The name "free running" comes from the fact that this circuit goes into action as soon as power is supplied. Both tubes begin to conduct, and capacitors C1 and C2 begin to charge, with their plate connections positive and their grid connections negative. Initially, the plate currents of the tubes

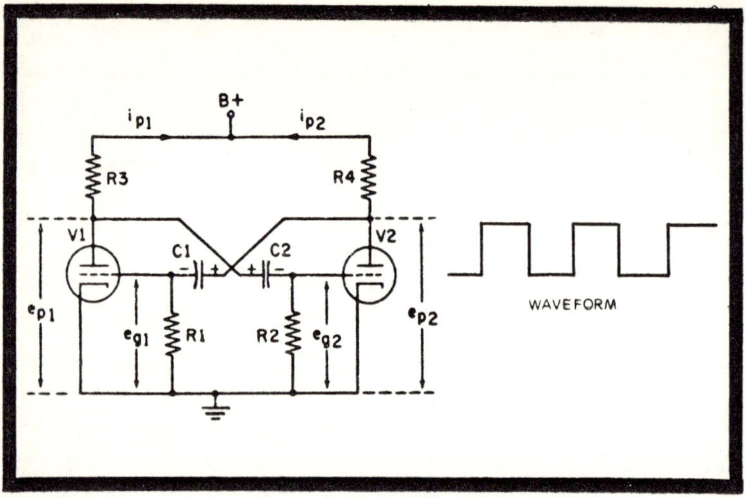

Fig. 4-16. Symmetrical multivibrator.

are nearly equal; however, there is always a slight imbalance in some of the circuit components. This imbalance will cause one of the tubes to begin conducting heavier than the other. When this happens, one plate voltage goes lower than that of the other, feeding back a negative signal to the grid of the lightly conducting tube. This signal is in phase with the grid signal and a regenerative situation occurs, which soon causes one tube to cut off and one tube to conduct heavily. This switching action occurs very rapidly, and it is followed by a longer period for which the tubes remain in their respective states.

In this example, assume that V1 in Fig. 4-16 begins to conduct heavier than V2. Plate voltage at V1 will begin to decrease, and C2 will begin to discharge from the grid connection to V2, through R2, up through V1, and back to the positive plate of C2. This discharge will cause a negative potential at the grid of V2, which, in turn, causes a decrease in the current through V2. The increase in plate voltage that results at V2 will cause C1 to charge through R4, the power supply, back to ground, and up through R1. In this manner, the charging current for C1 will increase the positive signal at the grid of V1, and V1 conducts heavier. This action is regenerative and results in cutting off V2. The circuit will remain in this state until a sufficient time has elapsed for C2 to discharge.

With V2 cut off, C2 discharges through R2, V1, and back to the opposite side of C2. As the charge across C2 decreases, the current through R2 decreases until a point is reached at which the voltage drop across R2 is no longer large enough to hold V2 cut off. At that point, V2 starts to conduct and its plate voltage drops. As it does, C1 starts to discharge through R1 and V2. The resulting current through R1 reduces conduction in V1, causing the plate voltage there to rise. Capacitor C2 starts to charge to that voltage by drawing current up across R2. This positive-going signal at the grid of V2 aids its conduction. Again the action is regenerative, but it results this time in V2 conducting and V1 being cut off.

The circuit will continue to operate in this manner, switching states at a frequency determined by the time constants of R1C1 and R2C2. As a result, a square wave is generated at the plate of each tube which can be capacitively coupled to an output circuit, where it is shaped into the sawtooth waveform necessary to drive the horizontal deflection plates of the crt.

The frequency and symmetry of the square wave available at the plate of V1 or V2 is a function of the RC time constants in the grid circuit. The longer these time constants, the lower the frequency. When the time constants are equal, the square wave will be symmetrical, as in Fig. 4-16. The symmetrical output furnishes a waveshape in which the half-cycles are equal in time. A more convenient waveshape for a sweep generator would be one in which a short retrace period is available along with a much greater period for the trace. A simplified schematic of a circuit that will furnish such an asymmetrical output is shown in Fig. 4-17.

The schematic is essentially the same as the one in Fig. 4-16, with one exception: R1 has been replaced with a potentiometer. If the resistance in the grid circuit of V1 is decreased

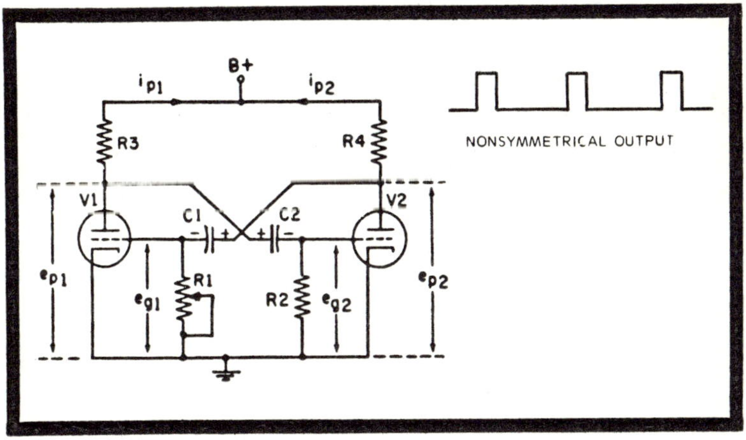

Fig. 4-17. Asymmetrical multivibrator.

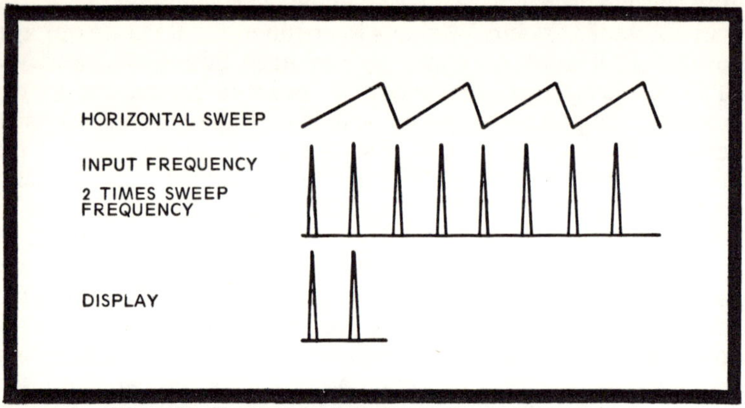

Fig. 4-18. Sweep and input signal synchronization.

to a point where the time constant of R2 and C2 is larger than that of R1 and C1, the waveshape will no longer be symmetrical. Instead, the decreased RC time constant in the grid circuit of V1 will cause that tube to be cut off for a shorter period of time, causing the asymmetrical output shown.

Syncing the Sweep

Using the RC-controlled multivibrator as the sweep generator has one major disadvantage. Even with the best capacitors and resistors used, the stability of the multivibrator is not good over a long period of time. As long as the oscillator is stable and does not drift and the vertical signal is in phase with the horizontal sweep, the sweep and signal will stay in step and a steady display will result (Fig. 4-18). But even a slight drift in the multivibrator frequency or input frequency will upset the stability. Therefore, another modification of the basic scope is necessary.

When a multivibrator is used in an application of this type, it usually is synchronized with some frequency that forces the period of the multivibrator oscillation to be exactly the same as that of the synchronizing frequency. When a synchronizing signal is used, the multivibrator is said to be **driven**. Sine waves or pulse voltages are generally used for the sync signal, although waveforms of almost any shape can be used. The synchronizing signal may be injected in series with the cathode circuit, in the plate circuit, or in the grid circuit.

In the example in Fig. 4-19, assume that the multivibrator starts with V1 conducting and V2 cut off. This is shown by the fact that from t1 to t2 in the voltage graph the grid of V1 is shown cut off and the plate voltage is low. At t2, the normal switching action of the circuit is initiated, the grid of V1 is driven far below cutoff, and the plate voltage rises toward the positive supply potential. Between t2 and t3, the cutoff condition of V1 is determined by the R1C1 discharge time constant. At t3, the grid voltage of V1 reaches cutoff, and normal switching action is initiated. At t4, the sync signal is applied to the grid of V1.

At this time, C1 has completed its charge and the grid potential is zero. At first glance it would appear that between times t4 and t5 the positive-going sync voltage would cause the grid to assume a positive potential. However, due to grid limiting, the grid voltage remains at about zero. This grid-limiting action can be explained by Fig. 4-19. As the sync voltage goes positive, grid current is drawn and it flows from the grid through the source of the sync signal, to ground, back to the cathode of V1, and from cathode to grid. Since the output impedance of the sync signal source is very large in comparison to the grid-to-cathode resistance of the conducting

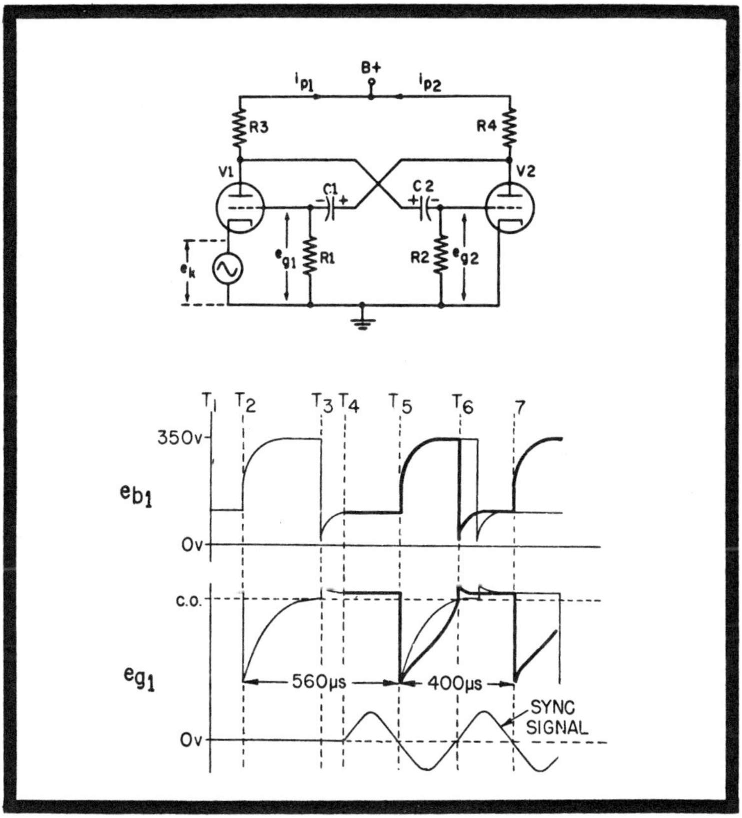

Fig. 4-19. Synchronized multivibrator.

tube, most of the sync voltage is dropped across the signal source on the positive half-cycle, and only a very small percentage of the positive voltage appears between the grid and cathode. For practical purposes, the grid waveform remains at zero during this half-cycle of the sync voltage.

The synchronizing voltage passes through zero and begins its negative half-cycle at time t5. During this negative signal, no grid current flows, and all of the sync voltage appears between the grid and cathode, adding to the voltage drop across R1. The resultant grid waveform is the sum of these two voltages; therefore, at t5, when the sync voltage becomes slightly negative, the grid of V1 takes on a small negative potential. This action decreases the plate current in V1 and starts the switching action.

The heavy trace in the figure represents the waveform due to the addition of the sync voltage, while the lighter trace shows the waveform as it would be with no sync voltage applied. Notice that from t5 on, the light and heavy traces are not coincident. Due to the effect of the negative half-cycle of the sync voltage, the grid is caused to approach cutoff at a much steeper angle than before.

At time t6, the sync signal passes through zero and begins a positive alternation. This positive sync signal rapidly forces the grid potential above cutoff and initiates the switching action before it would have occurred in the absence of a sync voltage. Between t6 and t7, the action is the same as between t4 and t5. At t7, the sync signal again starts a negative half-cycle, causing a negative signal at the grid, decreasing the plate voltage, and starting the switching action.

From this, it can be seen that shortly after the application of the synchronizing signal the period of the multivibrator is forced to be the same as the period of the sync signal. Also, use of a sine wave as a sync signal allows the sync signal to control both the conduction and cutoff times of the triggered tube. Although multivibrators are sometimes synchronized by sine-wave voltages, a more accurate synchronization is obtained by use of pulse voltages.

Basic Repetitive Sweep Circuit

A simplified schematic of an actual repetitive sweep circuit is shown in Fig. 4-20. The interstage-coupling network composed of R2, S1, and the capacitors selects the frequency of operation of the multivibrator and thus controls the horizontal sweep time of the oscilloscope. By switching different capacitors into the circuit, different RC time constants are formed and the sweep time is altered. Capacitor C1 is the sawtooth-forming component, having a long charging time through R4 and the power supply and a short discharge constant through R1 and V1.

At time t1 in the voltage graph, the multivibrator has just completed the retrace and is starting the trace portion of the cycle, as indicated by the fact that the plate voltage of V1 (the voltage across C1) is at its minimum value. Tube V2 is conducting, and the resultant current causes a slight voltage drop across resistor R1. The plate voltage of V2 is near minimum, causing the selected capacitor to discharge through grid resistor R2. This applies a large negative potential to the grid of V1, as shown by the waveform. Since V1 is cut off at this time, capacitor C1 will begin charging toward the power supply voltage. However, the long charging time of C1 allows the capacitor to charge to only 20 percent of its maximum charge by the time t2 occurs. This insures that the most linear portion of the sawtooth is used as the sweep voltage.

As C2 discharges between times t1 and t2, the grid voltage of V1 (e_{R2}) decreases toward the cutoff value of V1. At the same time, plate current in V2 increases slightly, causing a drop in plate voltage. This is because a portion of the bias for V2 is derived from the discharge of C2 through R1. As C2 discharges the bias on V2 decreases, resulting in a gradual increase in plate current.

At time t2, the voltage across R2 has reached the point of conduction for V1. At this instant, an action similar to the switching action of a normal multivibrator takes place, and V2

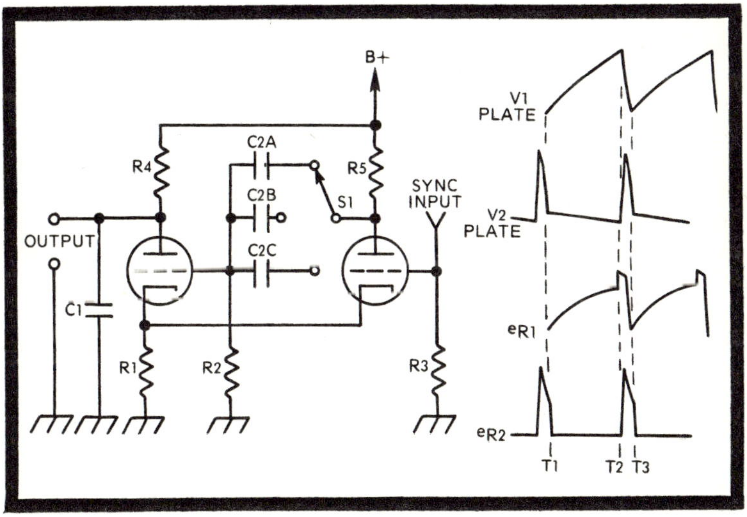

Fig. 4-20. A simplified sweep multivibrator.

is cut off. The rounding of the leading edge of the voltage waveform across R2 is caused by the charging current for C2 flowing through R5. The charging current of C2 causes a positive potential on the grid of V1, which increases the plate current and lowers the plate voltage. This decrease in plate voltage at V1 causes C1 to discharge rapidly through the relatively low resistance of R1 and the plate resistance of V1. At time t2, the current through R1 is made up of V1 plate current, V2 plate current, charging current for C2, and the discharging current of C1. These currents will cause a sharp rise in the voltage across cathode resistor R1 (e_{R1}), applying a positive spike to the cathodes of V1 and V2. As C2 charges, the positive grid voltage decreases. While C1 discharges, the resultant current through R1 decreases, as shown by the decrease in e_{R1} between t2 and t3. As the cathode voltage decreases, the plate voltage of V2 decreases until, at t3, it falls below the potential on the positive plate of C2. This causes C2 to begin to discharge and initiate the cutoff of V1.

This action completes one cycle of operation. The time between t1 and t2 is the sweep portion of the cycle, and the time between t2 and t3 is the retrace part. Now, let's look at the complete schematic of the horizontal sweep circuit in an actual oscilloscope.

Practical Sweep Circuit

The complete schematic of a repetitive horizontal sweep circuit is given in Fig. 4-21. A 6-position switch, S104, the **coarse frequency adjust** control, provides an operating range from about 3 Hz to 50 kHz. This range of frequencies is controlled by capacitors C125 through C131. These capacitors act as sawtooth-generating capacitors for V108B and as coupling capacitors between the two stages. With the **coarse frequency** switch, S104, set in position 6, capacitor C130 is connected between the plate of V108B and ground, where it functions as a sawtooth capacitor. Capacitor C131 functions as the coupling capacitor in position 6. However, when S104 is turned to position 5, C130 (formerly the sawtooth capacitor) now becomes the coupling capacitor. In this manner, S104 selects the frequency range of the multivibrator.

Adjusting the operating frequency within the limits of a given range is accomplished by use of the **vernier frequency** control. This control consists of a potentiometer—R158A and R158B, ganged to a single shaft. Potentiometer R158A affects the operating frequency by varying the plate voltage of V108B, while R158B affects the operating frequency by varying the RC time constant of the coupling network. The sawtooth output of the V108B is applied to a voltage divider composed of R155 and R125. The ratio of these resistances is approximately

10:1, so the sawtooth waveform developed across R125 will be one-tenth the amplitude of the waveform at the plate of V108B. Capacitors C142, C123, and C113 are frequency-compensating components, while C124 functions as a coupling capacitor. Resistor R159 is the grid-current-limiting resistor which provides a sharp output pulse.

The sync pulse amplifier, V101, receives the sync pulse from one of three sources, as selected by S103. In the **int** (internal) position, a portion of the vertical input signal is used as the sync pulse; in the **ext** (external) position, an external pulse is used. A 60 Hz sine-wave signal from the power source can also be selected as the synchronizing voltage.

The output from the **sync selector** switch is coupled by C121 to the grid of V101, with R151 acting as the grid resistor. Resistor R153 is the cathode-biasing resistor. Plate voltage is furnished by the voltage divider consisting of R154 and R150. With a signal applied to the grid of V101, the high end of the **locking** control, R154, will be receiving signals developed at the plate of this tube, and the low end of the **locking** control will be receiving signals from the cathode. When this control is at approximately the center of its rotation, there is no signal, since the center of the control is grounded. If this control is

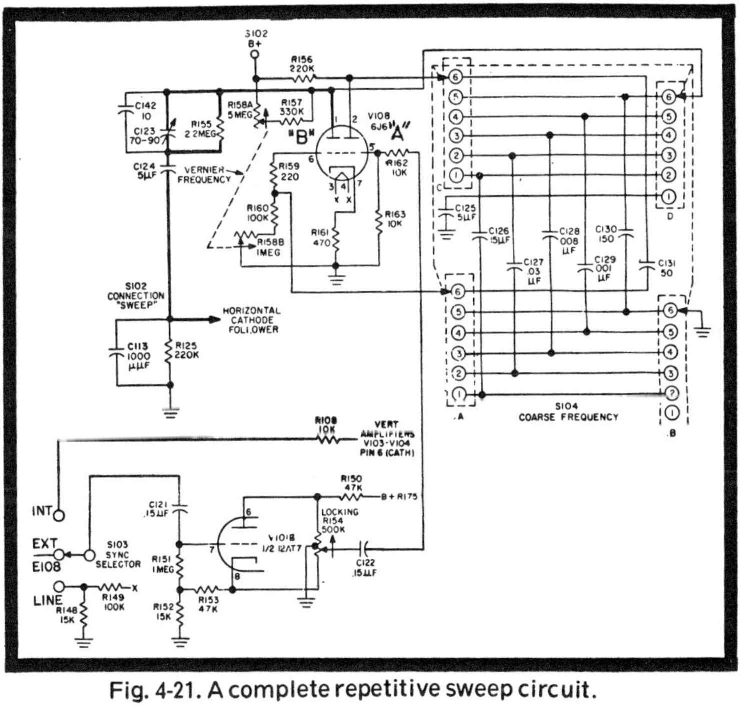

Fig. 4-21. A complete repetitive sweep circuit.

operated toward the plate side of R154, a locking voltage is obtained which is out of phase with the signal applied to the grid. If the control is advanced toward the cathode side of R154, the locking voltage applied to the sweep circuit multivibrator is in phase with the synchronizing signal on the grid.

The function of the locking control is to lock the sweep either in phase or out of phase with the synchronizing voltage. The amplitude of the output signal increases as the arm of R154 is moved away from the center position, providing a method of selecting the amplitude as well as the polarity of the output. From the wiper of R154, the sync signal is coupled to the sweep multivibrator through coupling capacitor C122.

Triggered Sweep

The repetitive sweep oscilloscope works well enough in most servicing applications. However, it is somewhat limited when accurate displays of high-frequency or very narrow pulse signals are necessary. A much more stable display can be achieved with a triggered-sweep presentation. The triggered-sweep circuit also offers the advantage of being able to observe expanded displays of a single cycle of the input waveform. Due to the advantages of this type of oscilloscope, many service shops are acquiring these instruments.

Do not confuse the operation of the triggered-sweep scope with that of the repetitive sweep scope in the internal sync mode. In both instruments, a portion of the vertical input signal is used to control the horizontal sweep. However, the repetitive sweep version uses the pulse as a sync voltage, while the same waveform in triggered-sweep scopes actually triggers the sweep circuits. No horizontal sweep is present in the triggered-sweep scope until a trigger pulse is applied. The monostable multivibrator in Fig. 4-22 can be used to explain the operation of triggered-sweep circuits.

This circuit has only one stable state. It remains in that state until triggered by an external signal, at which time it shifts its state, remaining in the new condition for a time determined by the value of the circuit components. At the end of this time, the circuit returns to its original condition. Although the sweep circuit of an actual scope differs somewhat from the simplified example, it is useful for our purposes here.

In the quiescent state, transistor Q2 is saturated by the positive voltage applied to the base through resistor R4. Transistor Q1 is cut off by reverse bias furnished by the negative voltage applied to the base through R2. The output voltage at this time is nearly zero, due to the low collector voltage at Q2. When an external positive signal is applied to

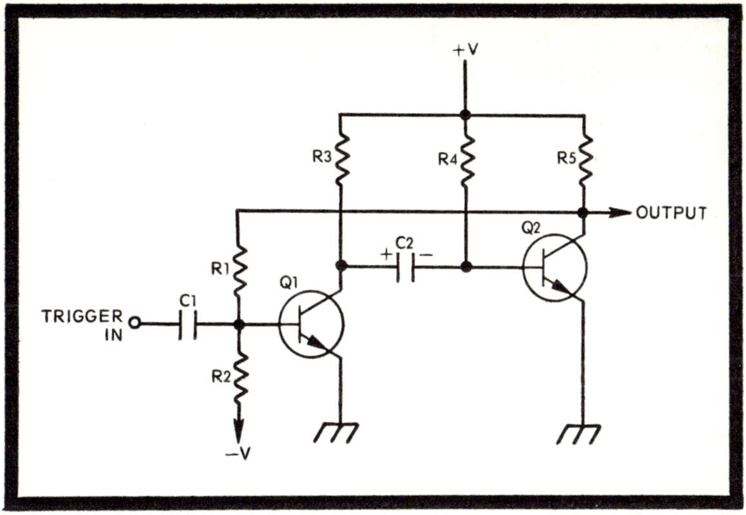

Fig. 4-22. A triggered-sweep circuit.

the base of Q1 through C1, it will tend to turn Q1 on. As the current through Q1 increases, the collector voltage decreases. The negative-going collector signal is coupled to the base of Q2 by C2, tending to cut off Q2. As collector voltage starts to increase, R1 feeds back an in-phase signal to the base of Q1, and regeneration occurs. As a result, Q1 is quickly driven into saturation and Q2 is cut off by the occurrence of the trigger.

The circuit is now in a quasi-stable state in which it remains as long as Q2 remains cut off. As C2 assumes a positive charge at the junction with R4, the base of Q2 rises toward the conduction level. When Q2 begins to conduct, feedback through R1 causes Q1 to cut off, and we're back to the stable state until another trigger arrives.

The output of this circuit is taken from the collector of Q2, and can be shaped into a sawtooth by an RC circuit. Sweep time, or the period of the output, is determined by the value of C2 and, as in our previous example, a switching arrangement can be used to provide several operating ranges. The sync signal furnished to this multivibrator should be a pulse voltage for best results. Since this signal is taken from the vertical input, it could have almost any shape, and for that reason, the sweep stage is usually preceded by a sync-shaping circuit.

The use of this type of sweep circuit allows the horizontal sweep of an oscilloscope to be triggered by the same waveform that is to be displayed. This insures perfect sync between the input signal and horizontal sweep, achieving a very stable display.

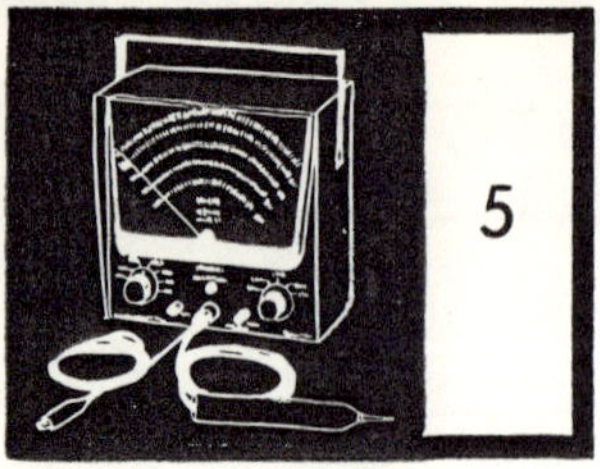

5 An Introduction To Waveform Analysis

For the crt display to be of any value, you must be able to recognize and interpret various features about the waveforms being displayed. This takes considerable practice. To make consistently accurate interpretations, you must be thoroughly familiar with the scope as well as the characteristics of the circuit being tested. You must know how your scope will display various waveforms and what methods work best with your equipment. The examples and explanations given here will supply you with a basic knowledge of waveform characteristics...the experience, **you** must supply.

SINE WAVES

What is a sine-wave voltage? It's a voltage that varies as the sine of the angle...everybody knows that. But can you attach any meaning to the key positions on the waveform? In Fig. 5-1, the sine wave is plotted for one complete cycle (360 degrees); amplitude is plotted as a percentage of the peak (maximum) voltage. Notice that all four slopes of the sine wave are proportional. That is, from 0 to 90 degrees the slope is the same as that from 90 to 180, except that it's reversed.

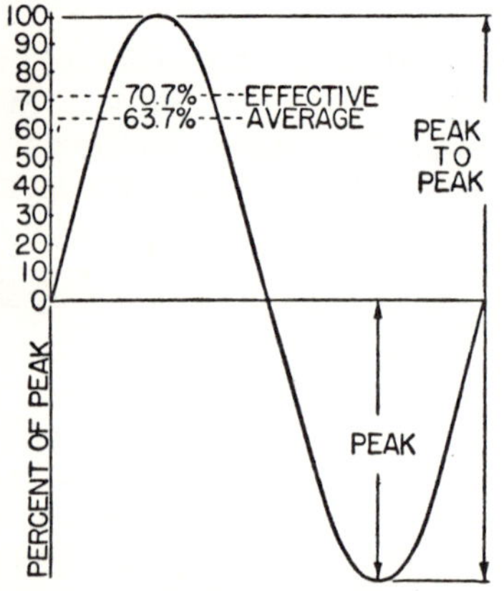

Fig. 5-1. The sine wave.

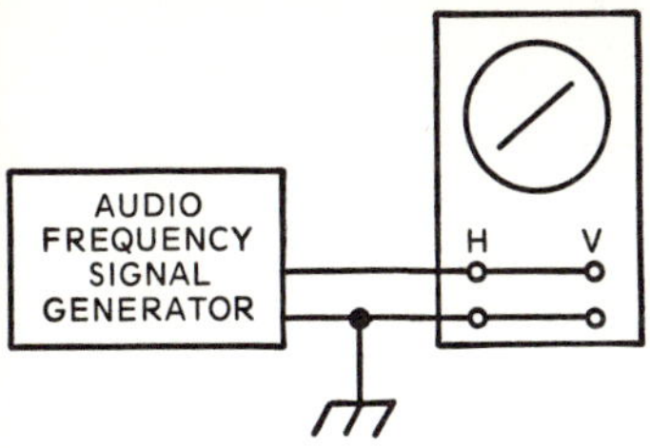

Fig. 5-2. Measuring phase shift of an oscilloscope's amplifier.

The same can be said for the other quadrants. If we examine the slope of the curve from 0 to 90 degrees, we can learn several facts about the entire cycle.

The slope of a sine wave is far from linear. If the wave were linear, it would pass through the 50-percent level at about 45 degrees. Instead, it passes the 50-percent level at 30 degrees, and by the time it reaches 45 degrees, it has obtained a level slightly above 70 percent. The rms value of a sine wave is 70.7 percent of the peak value. Notice how after passing the 60-degree point the rate of increase slows down to a point where, in the last 30 degrees (60-90 degrees), the gain in amplitude is only on the order of 17 percent; while in the first 30 degrees (0-30 degrees), the gain was 50 percent. These characteristics of the sine wave give it its familiar waveshape.

Phase Comparison

When an oscilloscope is used to determine the relative phase of two sine waves, it must be determined if the scope itself introduces any phase shift into the signal. Any phase shift in the oscilloscope circuits will introduce an error into the pattern being observed.

The amount of phase shift error present in the scope circuits can be determined by applying a single sine wave to both the horizontal and vertical inputs, as shown in Fig. 5-2. If there is no phase shift in either amplifier, the display will be as shown at the far left in Fig. 5-3. Other patterns are shown for various amounts of phase shift.

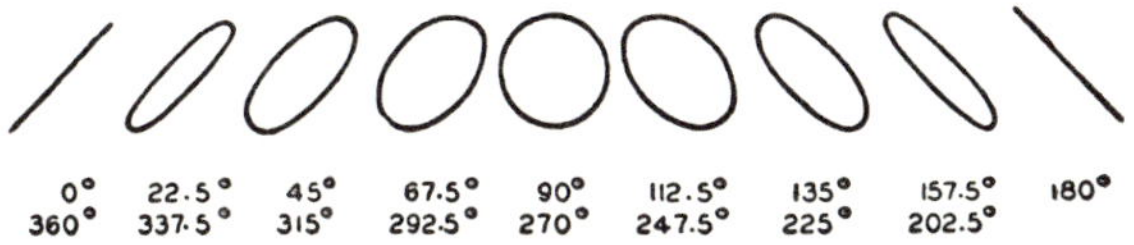

Fig. 5-3. Lissajous patterns indicating phase differences.

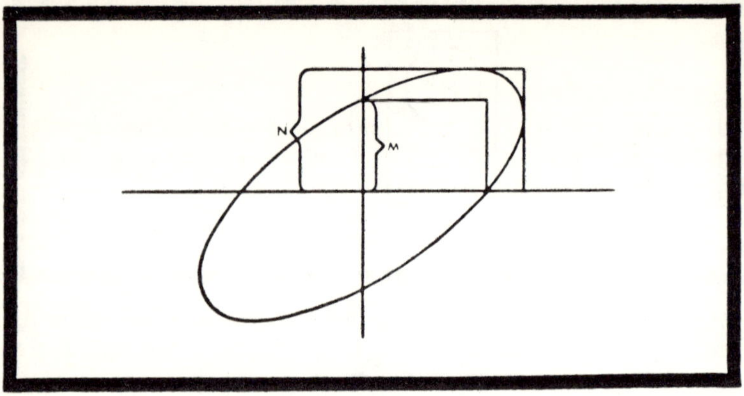

Fig. 5-4. Measuring phase shift.

To calculate the amount of phase shift more accurately, the display must be centered over the scope graticule as shown in Fig. 5-4. Measure the distance along the vertical axis to the point where the display intercepts the axis (m). Next, measure the distance along the vertical axis that corresponds to the maximum vertical deflection of the pattern (n). The angle of phase difference is found with the help of this formula:

$$\sin\theta = m / n$$

The angle θ is then found by referring to a sine table.

A similar procedure can be used to determine the phase angle between two sine waves. The same setup is used, but different sine wave signals of the same frequency are applied to the horizontal and vertical inputs. To be accurate, this

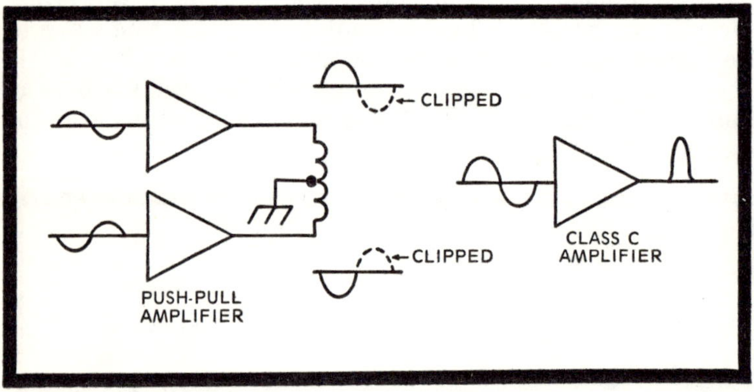

Fig. 5-5. Amplifiers having normally distorted outputs.

method must take into account the amount of distortion present in the oscilloscope itself.

Distortion in Sine Waves

From the technician's point of view, determining distortion is one of the most important uses of the oscilloscope. "Distortion" is a vague term, with many meanings. As used in a general way in connection with waveform observation, it simply means a departure from the desired waveshape; for example, a departure from the original sinusoidal form.

Some circuits have a built-in distortion that is desirable. Some classes of amplifiers, for example, normally clip the sine waves as shown in Fig. 5-5. Recognizing the difference between this and undesired distortion is necessary.

The circuit in Fig. 5-6 is an amplifier circuit that is biased class A. Recall that this class of amplifier normally delivers an output that is a true amplified version of the input. Any

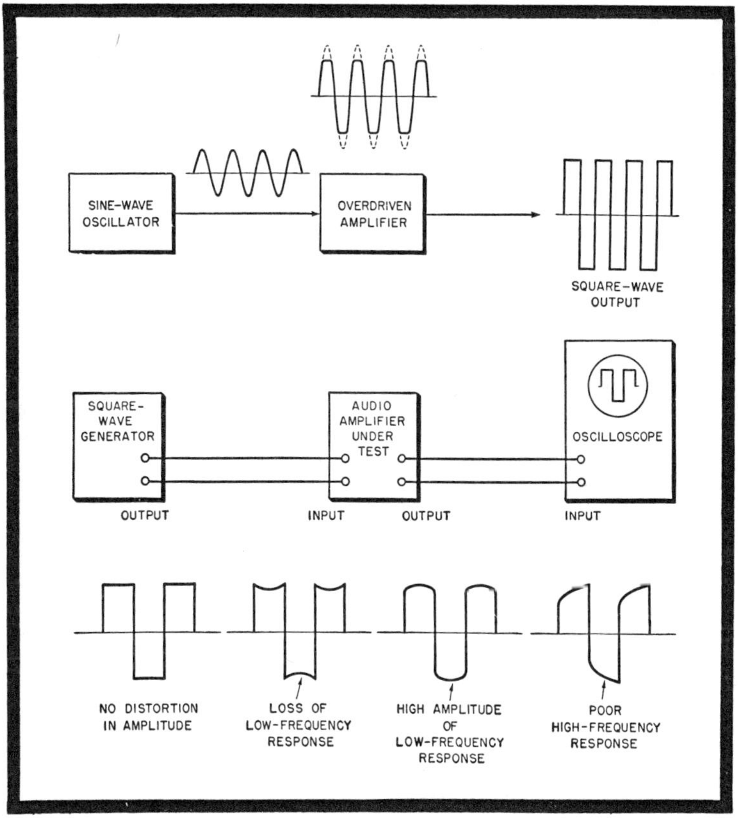

Fig. 5-6. Examples of distortion in class A amplifiers.

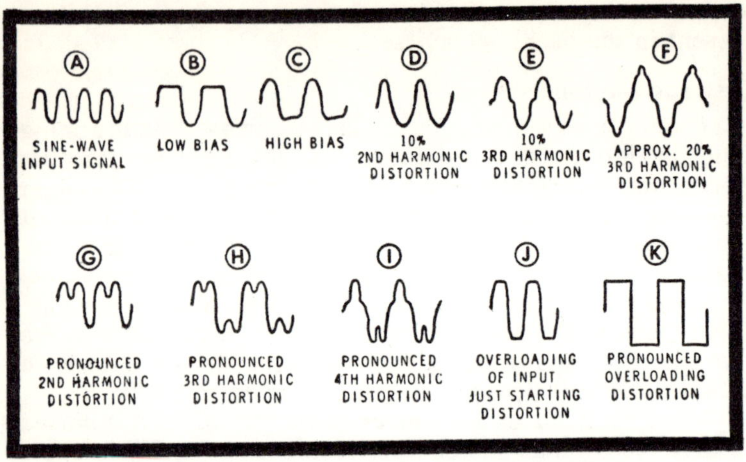

Fig. 5-7. Second- and third-harmonic distortion.

deviation from the input (other than expected amplification) is distortion. The examples of distortion shown here are caused by such conditions as insufficient bias, insufficient supply voltages, excessive base current, etc.

Impedance mismatch is often the cause of distortion. For example, a speaker that is not matched to the audio amplifier output will cause the waveform to be altered. In rf and i-f circuits, incorrect tuning can cause attenuation or over-emphasis of some component frequencies.

Other types of distortion can be traced to defective feedback circuits. **Negative feedback** is often used to extend the frequency band of an amplifier and increase the overall

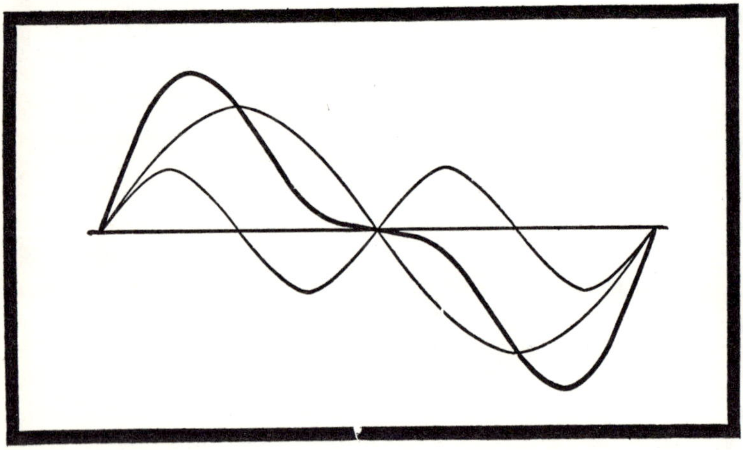

Fig. 5-8. How the distorted waveform is derived.

stability. **Positive** feedback is used in oscillator circuits to cause an amplifier circuit to oscillate. Defective feedback circuits can cause **ringing** (oscillations) in a conventional amplifier, and can dampen oscillations in an oscillator.

Distortion caused by incorrect operating voltages can be divided into two general categories. The first is where all operating voltages maintain their relative values with respect to each other. In this case, the waveform is usually not distorted, but its amplitude is changed. The other case is when the operating voltages fail to maintain their relative values. This usually will cause distortion.

Harmonics of the fundamental sine wave can cause severe distortion. Harmonic distortion is a result of combining the fundamental frequency with one or more harmonics. The waveshapes for various forms of harmonic distortion are illustrated in Fig. 5-7. How this distortion takes place is illustrated in Fig. 5-8, where the fundamental sine wave is combined with a second-harmonic wave.

Whether the distortion is caused by even or odd harmonics can be judged by a close observation of the resultant waveshape. (Refer to Fig. 5-9.) When odd harmonics (third, fifth, etc.) are combined with the fundamental, the positive and negative excursions are mirror images of each other. When only even harmonics are combined with the fundamental, the negative and positive half-cycles aren't mirror images.

Although it is fairly easy to recognize that harmonic distortion is present in a waveform, it is very difficult to determine the exact harmonic causing the distortion. For that reason nonsinusoidal waveforms are often used as test signals to observe distortion.

NONSINUSOIDAL WAVEFORMS

The sine wave represents the action of a swinging pendulum and other mechanical phenomena. It is a natural, or fundamental, waveform. Sine waves are present in various combinations in other waveforms, and no waveform composed of more than one frequency is a pure sine wave. Although clipping, harmonic, and other forms of distortion can

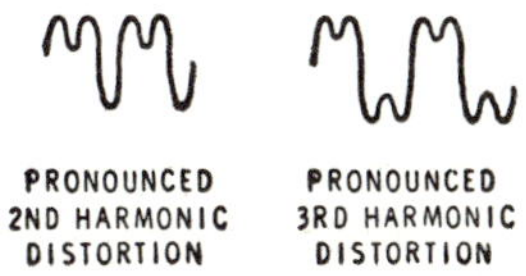

Fig. 5-9. Differentiating between odd and even harmonic distortion.

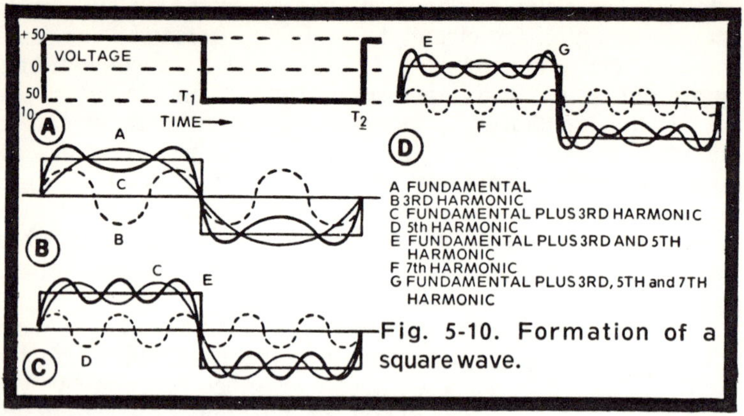

Fig. 5-10. Formation of a square wave.

sometimes be located using the sine wave, in some cases other waves offer a better picture of circuit operation.

Square Wave

The square wave is formed by combining a fundamental sine wave and a large number of odd harmonics. Although, as illustrated in Fig. 5-10, the addition of as few as three odd harmonics will produce a reasonable approximation of a square wave, a minimum of ten added harmonics are required to produce a useful square wave. Notice in the figure that as each successive harmonic is added, the sides of the resultant wave become steeper and the ripple at the top of the wave becomes smaller and higher in frequency. The leading and trailing edges of a square wave contain the high-frequency information and the flat top contains the low-frequency information. This is reasonable, since the rate of change is much greater in the leading and trailing edges than in the peaks.

In a square wave, the harmonics are in phase with the fundamental and inversely proportional to the harmonic order (the third harmonic contains one-**third** the amplitude of the fundamental, etc.). If a harmonic is not in phase or is of incorrect amplitude, the resultant square wave will be distorted. When it is observed with the aid of an oscilloscope, an experienced technician may be able to discern the kind of trouble or even the source of trouble in a circuit.

Other Waveforms

The **sawtooth** waveform is the result of combining both even and odd harmonics with the fundamental, as shown in Fig. 5-11.

The **trapezoidal** waveform can best be understood by regarding it as a combination of a sawtooth and a rectangular

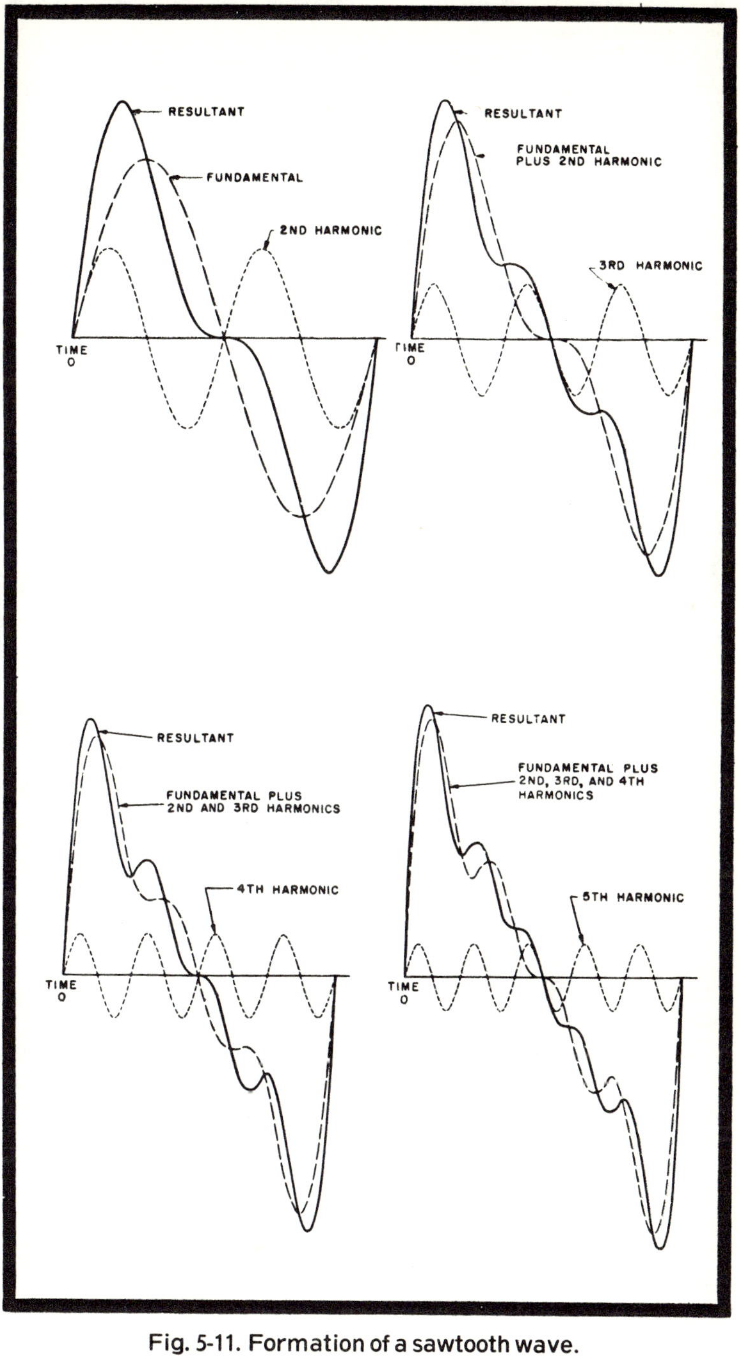

Fig. 5-11. Formation of a sawtooth wave.

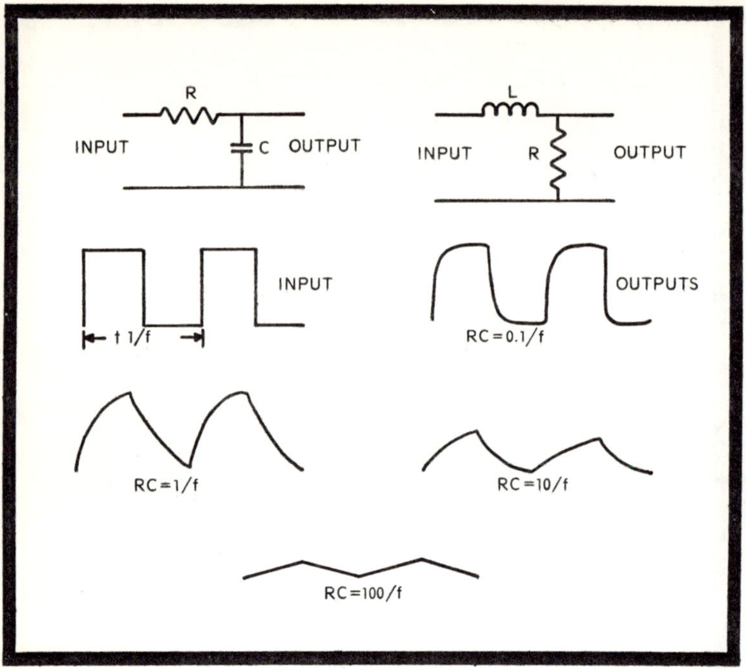

Fig. 5-12. Basic integrator.

waveform. Such waveforms are encountered frequently in the deflection circuits of cathode ray tubes using electromagnetic deflection. The trapezoidal waveform has the necessary characteristics to cause the linear change in current that is required in electromagnetic deflection systems.

Integrated Waveforms

Many nonsinusoidal waveforms can be resolved into their components by **integration**. Figure 5-12 illustrates two basic integrators. The exact effect of these circuits on a particular waveform will depend on the applied waveform and the time constant of the integrator (RC or L / R). For example, if the input is a steady dc voltage, a steady dc voltage output of the same value will be obtained. However, if the input is in the form of a series of narrow, widely separated pulses, the output voltage will be only a fraction of the peak input. How the resistance of the RC version affects the output waveform is shown in Fig. 5-12.

When a square wave is applied to an integrator with a long time constant, the output waveform will be a sawtooth of decreased amplitude. The shorter the time constant, the more closely the square wave is duplicated in the output.

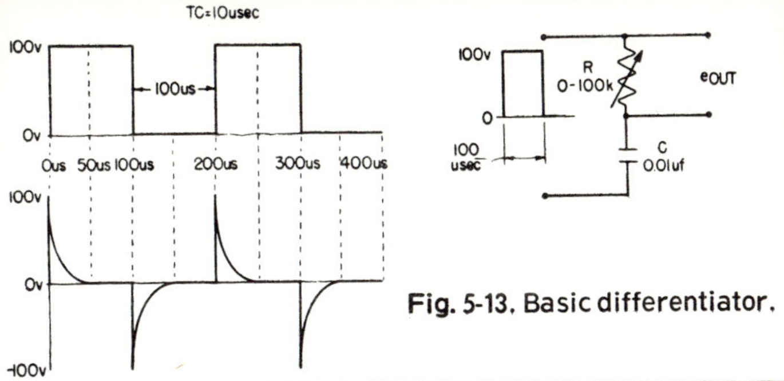

Fig. 5-13. Basic differentiator.

Differentiated Waveforms

The output of a differentiator is proportional to the rate of change of the input signal; a constant input produces no output. The differentiator is a type of high-pass filter; it extracts high-frequency sine-wave components, while the integrator removes low-frequency components. As in the integrator, the differentiator output is dependent upon the applied waveform and the time constant of the circuit.

In the circuit in Fig. 5-13, a rapid change in the input waveform will produce a narrow spike in the output, the peak amplitude being directly proportional to the rate of change of the input voltage. When a square wave is supplied as an input, sharp spikes are produced in the output as a result of the fast rate of change of the leading and trailing edges. The flat portions of the square wave will produce zero output, since they represent zero rate of change.

If a sawtooth waveform is applied to a differentiator with a short time constant, the output is a rectangular waveform. However, as the time constant is increased, the output gradually takes on the appearance of the input, as shown in Fig. 5-14.

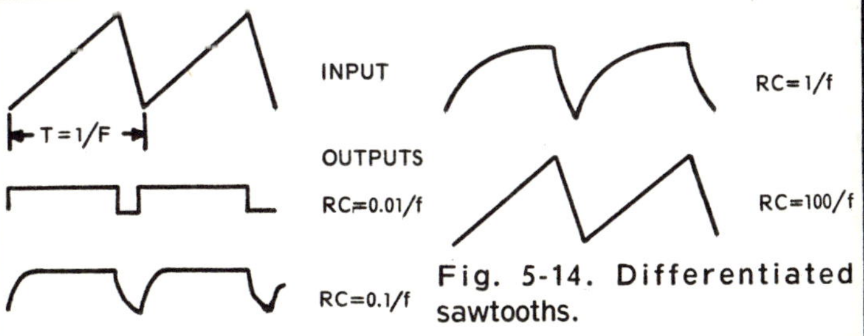

Fig. 5-14. Differentiated sawtooths.

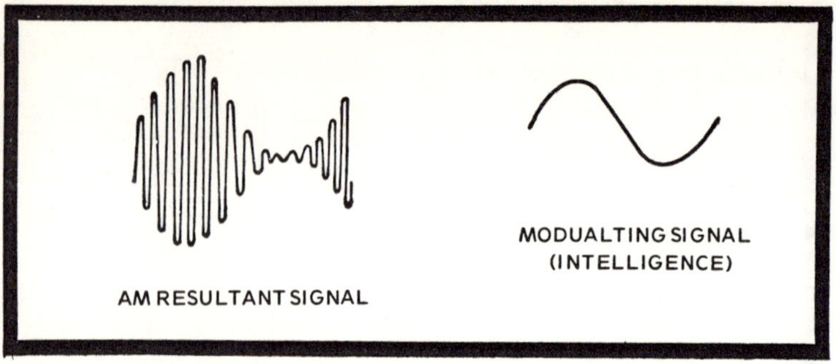

Fig. 5-15. An example of amplitude modulation.

Modulated Waveforms

The types of modulation to be considered here are AM and FM. Amplitude modulation, or AM, occurs when a constant-frequency carrier is altered in amplitude by an intelligence signal (Fig. 5-15). The rf signal spectrum after modulation is composed of the carrier frequency and two **sidebands**, one for each modulating frequency. For example, if the carrier frequency in Fig. 5-15 were 1000 kHz and the modulating wave were at 10 kHz, the resultant modulated signal would contain the original carrier, an upper sideband component resulting from the addition of the carrier and modulating waves (1000 kHz + 10 kHz = 1010 kHz), and a lower sideband obtained from the subtraction of the modulating frequency from the carrier (1000 kHz − 10 kHz = 990 kHz).

When the carrier is modulated by more than a single frequency, sidebands are generated at multiples of each modulating frequency, and the intelligence is contained within the sidebands created.

Frequency modulation occurs when a constant-amplitude carrier is shifted in frequency by a modulating signal. In this

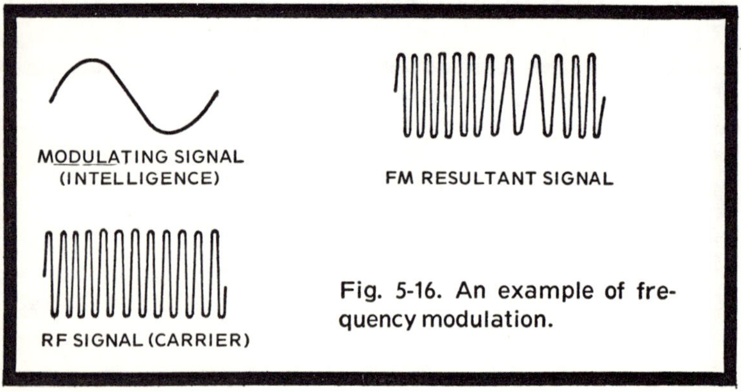

Fig. 5-16. An example of frequency modulation.

method of modulation, illustrated in Fig. 5-16, the modulating signal shifts the frequency of the carrier to represent the intelligence, and no amplitude change in the carrier occurs. Normally, the modulating signal alters the carrier so that the rate of frequency change in the carrier represents the frequency of the modulating signal, while the **amount** of frequency shift in the carrier represents the amplitude of the modulating wave.

Response Waveforms

Response waveforms are obtained by viewing a modulated waveform with an oscilloscope and a detector. The waveshape can be useful in determining just how a particular circuit will respond to various rf signals.

A response curve is a graph of the relationship between the output voltage and frequency. This curve can indicate the amount of amplification of various frequencies applied to a particular circuit or component. (Refer to Fig. 5-17.) Here a signal generator is used with an EVM to plot the response of an amplifier. The generator initially is set to a frequency near or below the low-frequency end of the bandpass of the amplifier, and the output voltage is measured and plotted. After the first plot has been made, the frequency of the generator is increased while the amplitude of its output is held constant. Now another output measurement is made and plotted. This sequence is continued until the limit of the high-frequency response capability of the amplifier is reached. When the plotted points on the graph are connected, the result is a curve

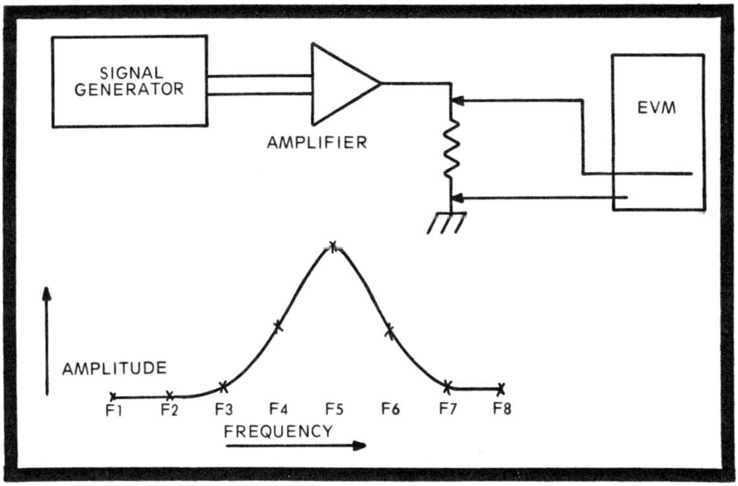

Fig. 5-17. Simplified example of frequency response testing.

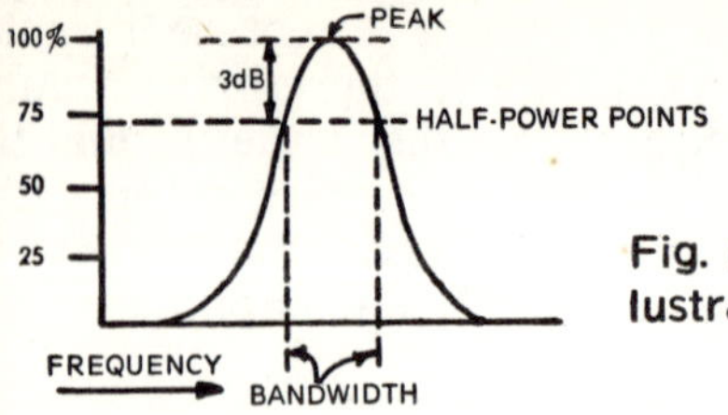

Fig. 5-18. A response curve illustrating circuit bandwidth.

that indicates the output level of the amplifier at any frequency in the passband (Fig. 5-18).

In the figure, the bandwidth is determined by the width of the curve at the **half-power points**, these points being 3 dB down from the maximum output, or peak, of the curve. Depending on the particular circuit and test equipment used, the curve may either be positive (as shown) or negative. Later, you will see how to obtain such a curve on a scope.

DISTORTION IN NONSINUSOIDAL WAVESHAPES

Although a sine-wave signal is useful in testing some amplifier circuits, in many cases this waveshape doesn't offer the best properties for analysis of circuit performance. Such things as clipping or harmonic distortion may be obvious from a simple observation with an oscilloscope when a sine wave is used as a test signal, but in cases of reduction in bandwidth and other frequency response conditions, the sine wave becomes a very hard signal to interpret.

Square-Wave Testing

Square waves are frequently used for fast and accurate determination of the overall operational characteristics of audio and video amplifiers. These waveforms can be used to determine the frequency and phase characteristics of an amplifier, because the square wave can be resolved into a series of frequencies with definite relationships. Figure 5-19 illustrates several amplifier waveforms that indicate certain defects.

One of the first visible signs of distortion in an amplifier will appear in the form of a phase shift. With a phase error of only 2 degrees, there will be as much as 10 percent tilt in the flat portion of the square wave. Very small phase errors can be observed with this technique when a low-frequency square-wave test signal is used.

Figure 5-19 illustrates the effects of low-frequency problems, again using a low-frequency square wave as the test signal. In a case of low-frequency loss near the fundamental frequency of the square wave, the dip in the top of the wave will be symmetrical. Combined low-frequency phase shift and response problems are indicated by the waveforms in Fig. 5-19.

Poor high-frequency response is also illustrated in Fig. 5-19. A gradual rise of the leading edge is caused by the combined action of high-frequency attenuation and phase shift. When the corners are slightly rounded and the rest of the waveform appears normal, phase lag is responsible. As the fundamental frequency is increased closer to the upper limits of the amplifier, the leading edge becomes more exponential. When a boost in high frequencies occurs, a spiked waveform will often result. Oscillation will cause fluctuations to be superimposed on the square wave as shown. In this case, look for regeneration in one of the stages. When resonance in an amplifier causes a specific frequency within the bandpass to be attenuated, a narrow dip will appear in the top of the square wave.

From these examples it is apparent that the square wave offers a very versatile and rapid means of waveform analysis of amplifiers. Although the rectangular wave is not used as often as the square wave as a test signal, an analysis of it is similar to that of the square wave, and it can be used for testing in the same way.

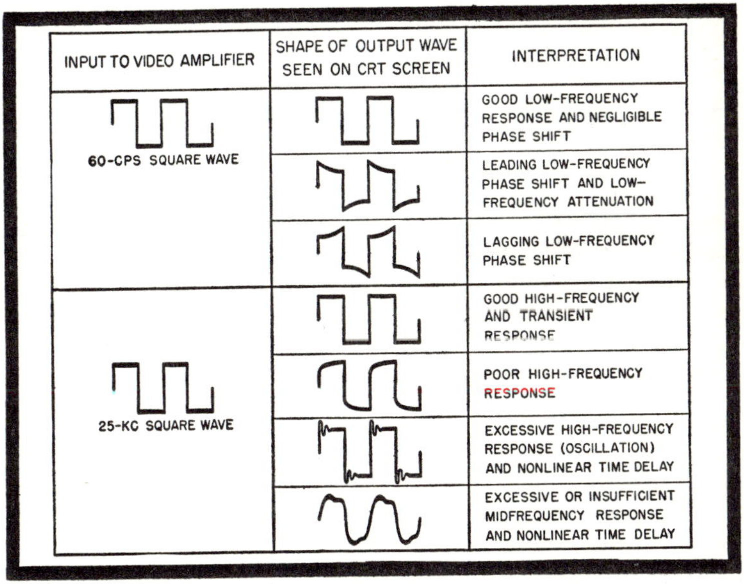

Fig. 5-19. Examples of square-wave distortion.

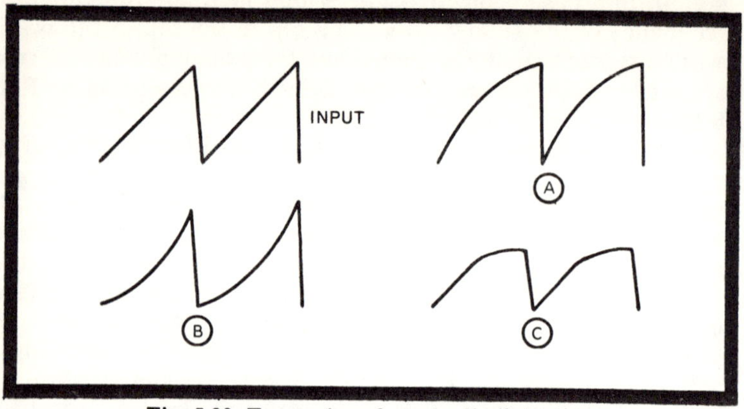

Fig. 5-20. Examples of sawtooth distortion.

Sawtooth Distortion

As discussed earlier, the sawtooth is usually generated by charging a capacitor through a resistance, using a high-voltage source. Some of the troubles that will distort this waveshape are leaky capacitors, increased or decreased values of capacitance or resistance, and changes in operating voltages. Figure 5-20 illustrates a few examples of sawtooth distortion. Example A can be caused by an amplifier drawing too much input current and allowing the capacitor to charge for a longer time. The distortion shown in example B is caused by incorrect operating potentials. Waveform C is produced by leaky capacitors, excessive bias, or excessive drive.

MEASUREMENT OF BASIC CHARACTERISTICS

In previous sections, we considered some of the general characteristics of waveforms that can be observed with an oscilloscope. In this discussion, we look at some actual measurements, using specific values.

Every waveform has some characteristics that are common to all waveforms. In the earlier chapters of this book, you saw how the EVM could be used to measure some of these values. In most instances, the oscilloscope gives a more detailed representation of circuit performance than the EVM does.

Voltage Measurement

Most service oscilloscopes have a provision for either ac or dc input. In the ac mode, a blocking capacitor is inserted in the input circuit of the vertical amplifier, allowing only ac signals to be coupled into the scope. When the dc mode is selected, the input circuits are wired directly into the vertical amplifier. This causes the displayed waveform to be a complex signal composed of both ac and dc components. If the input is a pure dc voltage, devoid of any ac signal, the oscilloscope display will be a straight horizontal line deflected from the zero position by an amount proportional to the amplitude of the dc voltage. In Fig. 5-21B, a positive dc input voltage is shown to cause the trace to be deflected upward. A negative dc voltage causes a downward deflection (Fig. 5-21D).

The effects of a complex input voltage on the scope display are illustrated in Fig. 5-22. With no signal applied to the vertical input of an oscilloscope, the display is a straight horizontal line. When the scope is set up for dc inputs and a complex waveform consisting of an ac signal riding on a dc level is applied to the vertical input, the display will be as shown at B. Here, a small sine wave is riding at some positive level. The scope trace is a sine wave displayed above the zero reference level. If the scope is shifted to the ac mode, the dc component is blocked by the input capacitor of the scope and only the ac component is displayed, as shown at C.

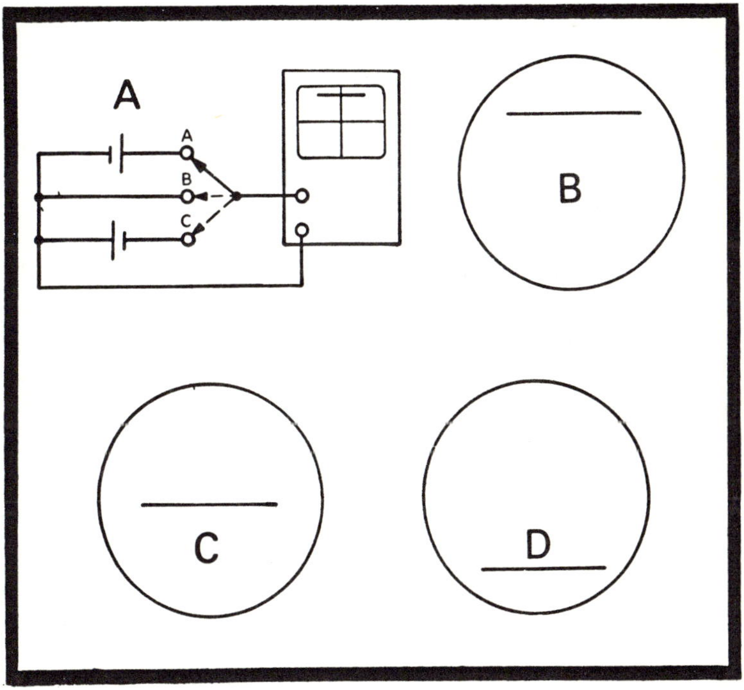

Fig. 5-21. Effects of dc voltage on scope trace.

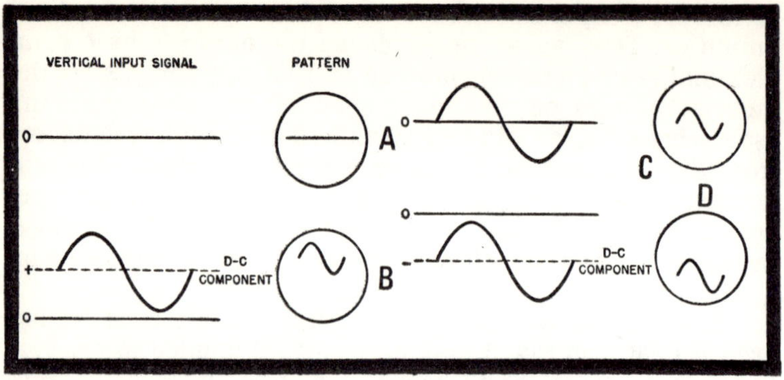

Fig. 5-22. Effects of ac and dc modes.

Voltages are measured with an oscilloscope by comparing the waveform to the **graticule** over the crt screen. This graticule is a scale divided into calibrated units. The number of units of deflection can be counted and the exact voltage computed. Later we will see how this voltage is computed on two different oscilloscopes. At this time, however, we will limit our computations to the number of units of deflection on the graticule.

In Fig. 5-23, an oscilloscope display of a complex waveform having both ac and dc components is illustrated. This waveform has a peak-to-peak ac component of 4 units, while the dc component is 1 unit. To eliminate the dc component, the oscilloscope can be shifted to the ac mode, and the display will be a sine wave of 4 units peak-to-peak, displayed about the zero reference level.

Although no direct display of circuit current can be made without the aid of special probes, it is possible to measure

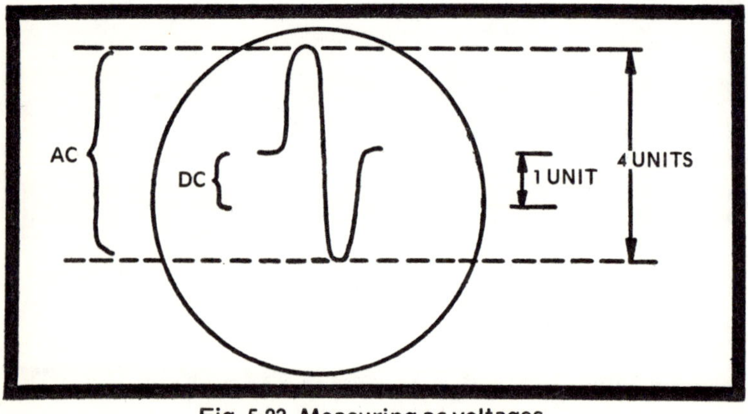

Fig. 5-23. Measuring ac voltages.

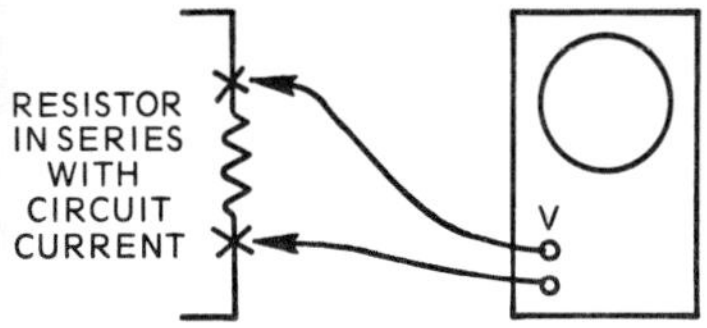

Fig. 5-24. Measuring current.

circuit current with an oscilloscope by making minor alterations to the circuit under test. Instead of connecting the oscilloscope in series with the current to be tested, a very small resistor is inserted. The leads of the oscilloscope are then connected across the resistor as shown in Fig. 5-24. This method will work in most circuits where the resistance added is not great enough to upset circuit operation. Actually, I seldom find a need to monitor circuit current with an oscilloscope in general service work, but this method has proven effective when that measurement is necessary.

Voltage measurement with the oscilloscope is an effective method of trouble isolation in complex signal-carrying circuits. By following the ac signals through the circuits and observing changes in amplitude and shape, many troubles can be pinpointed. Comparing input and output signal levels will reveal stage gain. This is a very fast method of isolating a faulty stage in a series of amplifiers. In order to measure voltage conveniently, the oscilloscope must be provided with calibrated vertical input circuits and a calibrated scale on the crt. These items are standard equipment on modern service scopes.

Frequency Measurements

Although no direct measurement of frequency can be made with an oscilloscope, the quantity can be calculated if the scope has a linear sweep. For these measurements, the sweep of the average repetitive sweep scope is not sufficiently linear; a triggered-sweep scope should be used. Triggered-sweep scopes are provided with a calibrated **horizontal frequency** control that allows the length of the horizontal sweep to represent precise intervals of time. By displaying the input waveform along this calibrated time base, the frequency can be computed. Figure 5-25 contains a table showing the frequency of the waveform for various periods.

To illustrate how frequency is measured with a triggered-sweep oscilloscope, assume that an ac signal is applied to the scope and the display in Fig. 5-25 results. The **horizontal sweep**

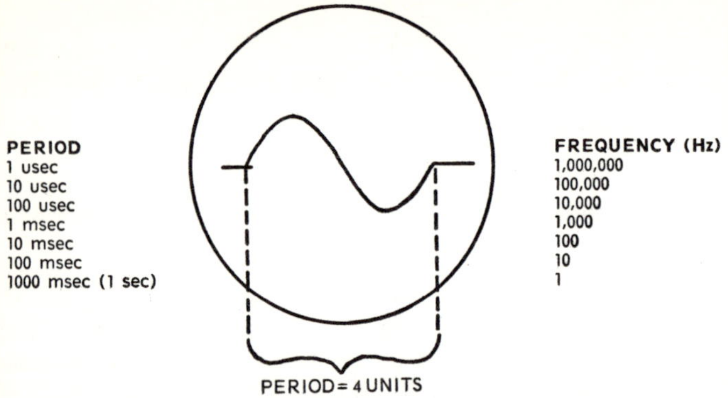

Fig. 5-25. Frequency measurement.

control causes each unit of the graticule to equal a specific period of time. A wide range of values can be selected on most scopes, allowing each division on the scale to represent various periods. Assume in this example that the control is at a position where each horizontal division represents 1 usec (one microsecond). This waveform requires exactly four divisions on the graticule to complete one cycle. The waveform, then, has a period of 4 usec. Since the period is the reciprocal of the frequency, the frequency is

$$f = 1/p = 1/(4 \times 10^{-6})$$

$$= 0.25 \times 10^{6} = 250{,}000 \text{ Hz} = 250 \text{ kHz}$$

In radar and similar pulse-operated equipment, it is sometimes necessary to make pulse width measurements. These measurements are similar to the period and frequency

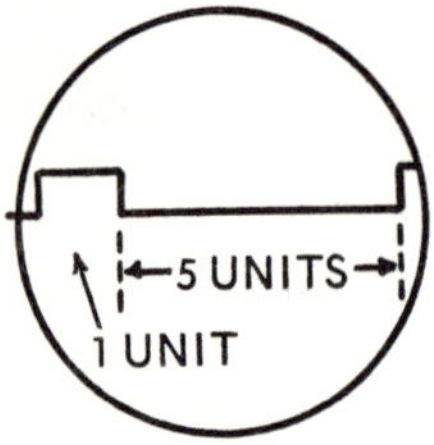

Fig. 5-26. Pulse width measurement.

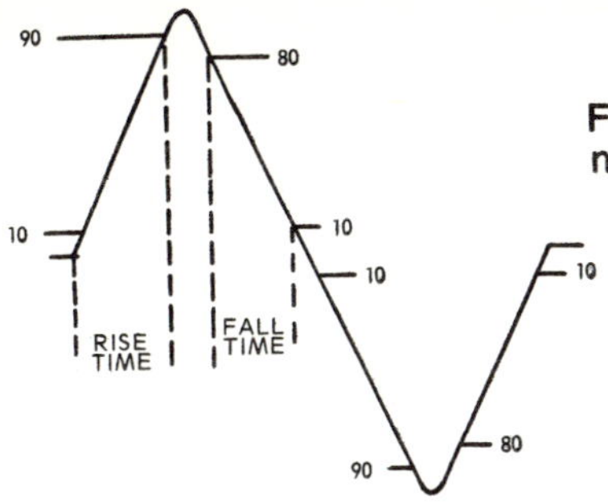

Fig. 5-27. Rise and fall time measurements.

measurements just discussed. Pulse waveforms are rectangular waveshapes having a very narrow period of time in which the level of the signal is raised (rise time) followed by a relatively long period of time in which the signal is steady. To measure pulse width, the waveform is displayed along the horizontal axis as shown in Fig. 5-26, and the number of units occupied by the pulse is noted. If the **horizontal sweep selector** remained in the **1 usec/div** (one microsecond per division) position, the waveform in the example would have a period of 6 usec, since the waveform covers six divisions per cycle, and the pulse width would be 1 usec.

Rise and Fall Time Measurements

Rise time is usually defined as the time it takes for the voltage to rise from 10 percent of its maximum value to 90 percent of its maximum value, as shown in Fig. 5-27. In some cases, the rise time will be taken from the 1 dB point to the 6 dB point. The "faster" the rise time, the shorter the amount of time it takes the voltage to rise from the 10 percent to the 90 percent point.

The **decay time**, or **fall time**, is usually measured between the points where voltage falls from 80 percent of its maximum amplitude to the 10 percent level. The smaller the decay time, the steeper the slope of the trailing edge.

6 The Wideband Oscilloscope

When referring to service-type oscilloscopes, it is common practice to use the description "wideband" or "narrowband." Basically, these terms have reference to the bandwidth limitations of the vertical amplifier. Although I have never seen an exact dividing line established between the two—a specific frequency at which a narrowband oscilloscope becomes a wideband scope—generally, the term "wideband" is used in connection with oscilloscopes with a vertical amplifier of sufficient bandwidth to observe the waveforms encountered in general service work. This usually ranges from dc up to the 5 MHz region. Scopes in this category usually have a fairly flat frequency response up to 4 or 5 MHz. In this discussion, we shall define as "wideband" those oscilloscopes having less than a 3 dB deviation from 0 to 4 MHz. These instruments are of the general service variety, used in radio and television receiver servicing.

The scopes found in most service shops are of the wideband category. Why is this true? The added bandwidth is sure to increase the cost.

Remember that the waveform presentation is a graphic plot of voltage against time. As the beam is swept across the screen by the time base, the vertical input signal voltage causes the beam to be deflected vertically. This results in a display that represents the "shape" of the voltage being observed. Recall from basic electronics that most amplifiers are designed to operate within a certain band of frequencies. When an amplifier is supplied a sine wave above the designed frequency, the overall signal gain of the amplifier is reduced. If frequency is steadily increased, eventually a point is reached where the signal is attenuated instead of amplified, thus rendering the amplifier useless in that frequency range. This reduction in amplification is one type of distortion.

This same drop in gain also has a detrimental effect on square waves. Recall that a square wave is composed of an infinite number of sine waves. When an amplifier is operated at such a high frequency that sine-wave attenuation occurs,

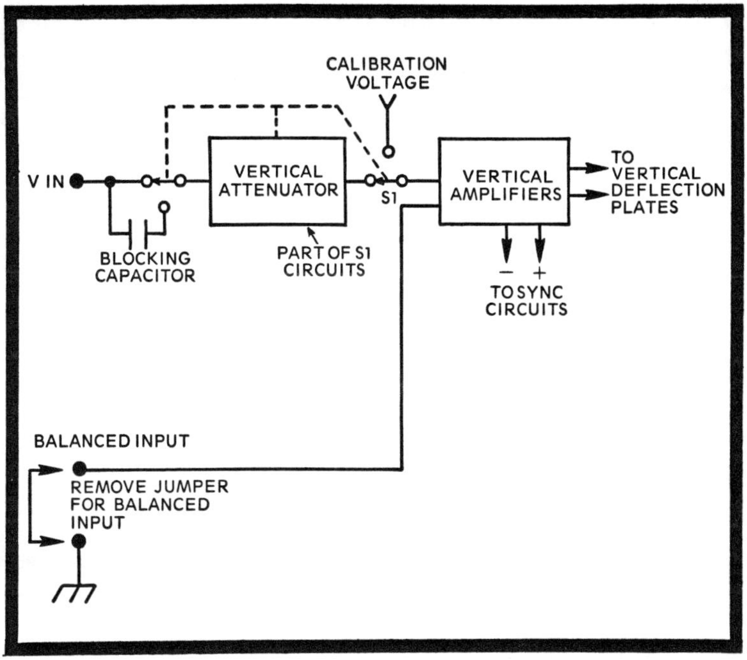

Fig. 6-1. EICO 460 vertical input circuits. (Courtesy EICO Electronic Instrument Co., Inc.)

the high-frequency components of the square wave surely will be attenuated. This causes the high-frequency parts of the display to be reduced in amplitude, rounding the edges of the square waves and producing a display that is not a true representation of the input signal.

Although some high-frequency attenuation can be tolerated, the more accurately the display represents the input signal, the better the technician can judge circuit operation. For this reason, most service oscilloscopes are wideband instruments.

THE EICO 460

One such wideband service oscilloscope is the EICO 460. This scope comes in either factory-wired or kit form, making the price well within the range of most hobbyists and technicians.

Vertical Circuits

A block diagram of the vertical circuits of the EICO 460 is shown in Fig. 6-1. From the **vertical input** jack, the signal is routed through the **vertical attenuator** switch and its

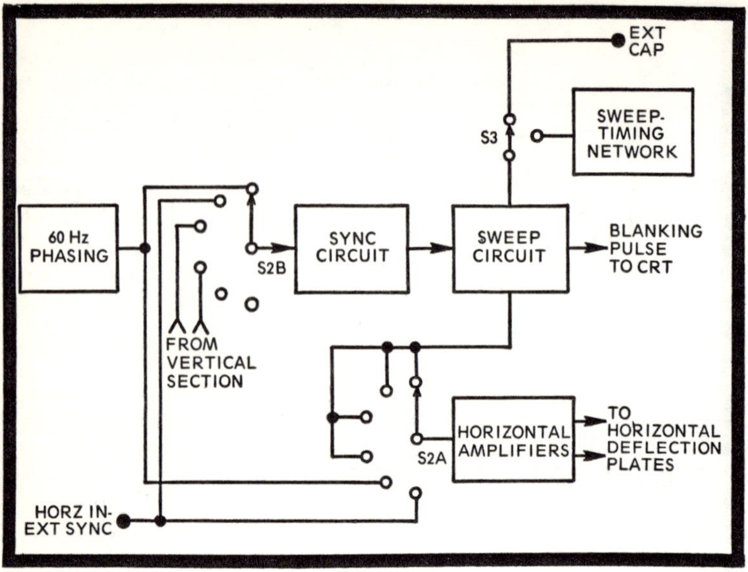

Fig. 6-2. EICO 460 horizontal circuits. (Courtesy EICO Electronic Instrument Co., Inc.)

associated circuitry to the vertical amplifiers. The attenuator also provides for ac coupling by means of a blocking capacitor.

Notice that three input jacks are available at the vertical input. One is the chassis ground; another, the conventional **vertical input** jack; and a third, the balanced-input jack. During normal operation, the probe is connected between the **vertical input** jack and ground, a shorting clip being used to tie the balanced input to ground. The balanced input feature allows any in-phase signal applied to both inputs to be rejected by the amplifiers. This will reduce considerably the amount of noise pickup in the test leads. When this method of waveform observation is being used, the balanced-input jack is left ungrounded and the signal to be observed is applied between the two vertical inputs. In this mode, there is no vertical attenuation and the scope operates at full vertical sensitivity.

The vertical amplifiers of the 460 give the scope a sensitivity of 25 mV per inch of vertical deflection, with a bandwidth of +1 to —3 dB from dc to 4.5 MHz. In the unbalanced mode, the maximum input voltage is 600V (dc plus peak ac), while in the balanced mode, the input is limited to 1V peak-to-peak.

Two outputs are taken from the vertical amplifiers. One, of course, feeds the vertical deflection plates of the crt. The second vertical output picks off a portion of the signal being observed and routes it to the sync circuits.

Horizontal and Sync Circuits

Figure 6-2 is a simplified block diagram of the horizontal deflection and sync circuits. In addition to the horizontal drive signal supplied to the crt, the horizontal output amplifier also furnishes a sawtooth waveform for test purposes to a jack on the front panel of the scope.

The amplifier is driven through the **horizontal selector** switch, S2A, by an input selected at that switch. Four **sync sweep** positions allow the horizontal amplifier to be controlled by the sweep circuit. Two other positions permit control of the horizontal amplifier by either an internal 60 Hz signal or by a signal from external equipment received at the **hor input ext sync** jack.

A multivibrator controlled by an RC network forms the sweep circuit. Switch S3 in one grid circuit of one multivibrator allows various RC combinations to be switched into the circuit to control the period of this stage. A sync amplifier and the other half of S2 complete the horizontal and sync circuits. A blanking pulse is sent from the sweep circuit to grid 2 of the crt to blank the screen during sweep retrace time.

We will refer back to these block diagrams a little la′er, when we discuss the front-panel operation of the scope.

CRT Circuits

The crt circuits employed with the EICO 460 are typical of conventional electrostatic deflection circuits. A high negative voltage is applied to the cathode of the crt as shown in Fig. 6-3. Grid 1 is supplied a negative potential somewhat higher than that of the cathode circuit. This potential can be varied by R72 to control the intensity of the electron beam. Another negative potential is supplied to grid 2 through R75 to focus the beam. Resistor R75 is adjusted coincidentally with R81 to achieve a sharp, uniform trace.

Notice the connection between the cathode and jack J9, marked "Z-axis." This jack provides for an input signal to be routed directly to the cathode of the crt for intensity modulation of the beam, which is useful in comparing the time relationship of two signals.

Operating Controls

The operating controls of the EICO 460 oscilloscope are shown in Fig. 6-4. In addition to controlling the illumination of

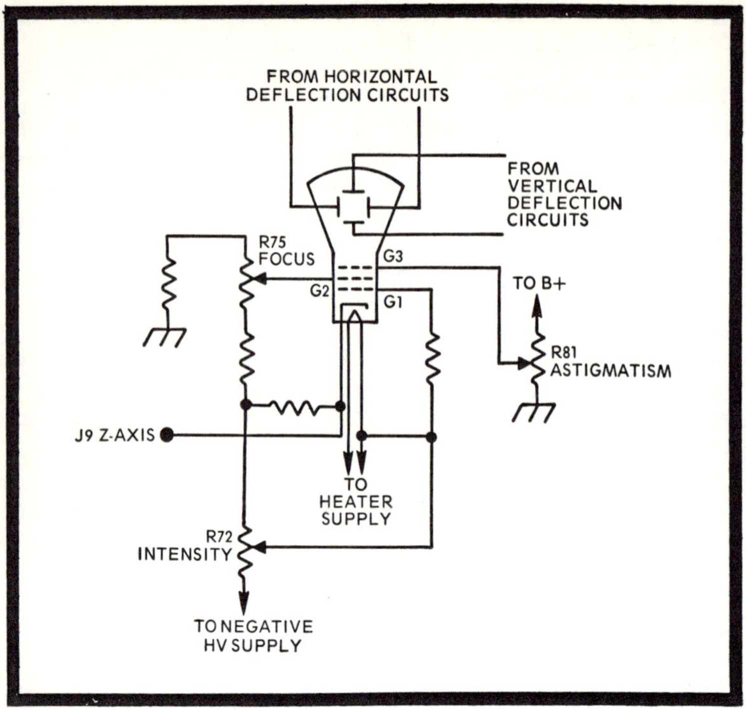

Fig. 6-3. Scope circuits. (Courtesy EICO Electronic Instrument Co., Inc.)

the scale graticule, the **scale light** control is ganged to a switch which supplies the necessary operating voltage to the scope. The **intensity** and **focus** controls, located just above the bezel, are adjusted for a beam appearance. Once a trace has been established, the **vertical** and **horizontal position** controls are used to position the trace at the center of the screen. With the beam positioning, intensity, and focusing controls set at the proper position, the scope is ready for operation.

The **vertical attenuator** switch provides a choice of no attenuation or one of three decade steps of frequency-compensated attenuation of the input signal. Each attenuation is obtainable with either direct coupling to the vertical amplifier, for positions under the **dc** arc; or capacitive coupling to the vertical amplifier, for positions under the **ac** arc. At either of the **1** positions, the amplitude of the input voltage is unattenuated. At the **10, 100,** or **1000** position, the amplitude is attenuated by that factor. When the switch is placed in the **cal** position, a 0.4V peak-to-peak sine wave at the power line frequency is supplied to the vertical amplifier.

Sverdrup | **Sverdrup Technology, Inc.**

From the Desk of . . .

DAMON BEYER

Need contract load
Sheets for DS-04
& DS-05 Bath 85-1471

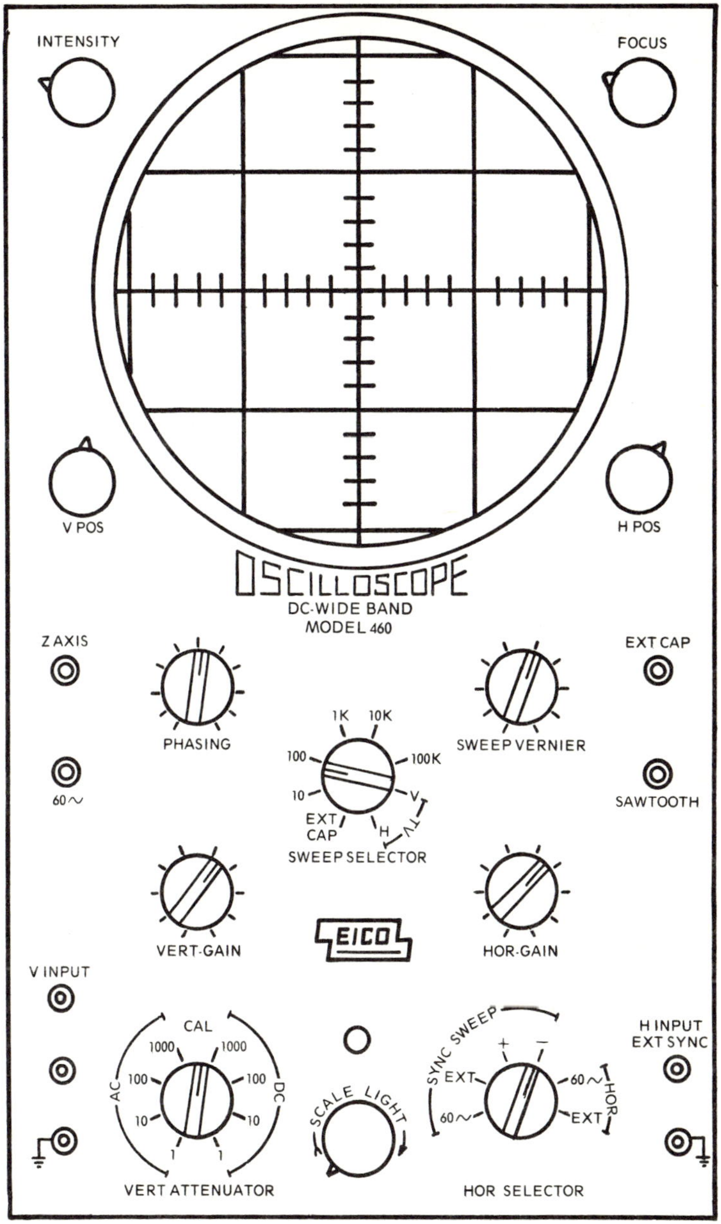

Fig. 6-4. Front-panels controls of the EICO 460 wideband oscilloscope. (Courtesy EICO Electronic Instrument Co., Inc.)

When the **vert** switch is set to a position under the **dc** arc, the dc component of any signal fed to the vertical amplifier will be amplified along with the ac component. This will appear on the screen as a vertical movement of the trace, the direction depending on the polarity of the dc.

The **vertical gain** control allows continuous adjustment of the vertical amplifier gain. It is used with the **vert attenuator** selector to adjust the trace height to the desired value.

The **hor selector** control combines the functions of sync selection for the internal sweep and input selection for the horizontal amplifier. At any of the four most counterclockwise positions (under the **sync-sweep** arc), the internal linear sweep is fed to the horizontal amplifier. These four positions permit selection of sync voltages for the sweep oscillator (see Fig. 6-2). At the + and — positions, the sync voltage is taken internally from the vertical amplifier. At the negative position, sweep trace flyback starts during the negative-going excursion of the input signal. At the positive position, the sweep trace flyback starts during the positive-going excursion.

When the **hor selector** is placed in the **ext** position, an external sync pulse applied between the **hor input-ext sync** terminal and ground is fed to the sweep oscillator. This feature is for use with generators or other devices which have sync pulse outputs. At the **60 Hz** position, an ac signal at the power line frequency is taken from the power supply and applied to the sweep oscillator to sync it at line frequency. The **phasing** control is effective in the **60 Hz sync** position, and permits phase adjustment of the 60 Hz sync voltage in order to shift the starting point of the display to any desired point on the waveform.

At either of the two most clockwise positions of the **hor selector** (under the **hor** arc), the sweep oscillator is disabled. In the **60 Hz hor** position, an ac signal of power line frequency is taken from the power supply and applied to the input of the horizontal amplifiers. The **phase** control determines the phase shift of the line voltage used for sinusoidal sweep. This permits the use of sweep generators that do not have phase-controlled 60 Hz sinusoidal sweep outputs.

Moving the **hor selector** to the **ext hor** position allows an external signal fed between the **hor input-ext sync** terminal and ground to be applied to the input of the horizontal amplifier. The **hor gain** control is effective in both the **60 Hz hor** and **ext hor** positions, controlling the gain of the horizontal amplifiers.

The **sweep frequency** band selector selects the frequency over which the **sweep vernier** control can be varied for frequency adjustment of the internal sweep. The numbers on either side of the marks are the approximate upper and lower limits of the band. The **ext cap** position is used when it is desired to obtain a linear sweep below 10 Hz, down to approximately 1 Hz. A 0.5 uF external capacitor must be connected between the **ext cap** and sawtooth terminals for this mode of operation. Larger value capacitors will give lower sweep frequencies.

The **V** and **H** positions under the **TV** arc are designed to eliminate the need for repeated adjustments of the **sweep vernier** when working with television receivers. If, with the **sweep selector** set at the **TV-V** position, the **sweep vernier** is set to a sweep frequency of 30 Hz (to display two full cycles of a 60 Hz signal), turning the **sweep selector** to **TV-H** will result in a sweep frequency of 7875 Hz, without readjustment of the **sweep vernier**.

SCOPE ACCESSORIES

To get the full value of the various operating features of any scope, it is necessary that the instrument be equipped with a proper assortment of accessories. Before we put the EICO 460 to work in some typical circuits, let's look at some of the accessories that can be used with it.

Direct Probe

In low-frequency or low-impedance circuits, where direct connection to the circuit under test is required to utilize maximum scope sensitivity and the added input capacity of the test leads will not distort the input signal, a **direct probe** may be used to connect the **vertical input** terminal of the scope to the test point. Such a probe is illustrated in Fig. 6-5.

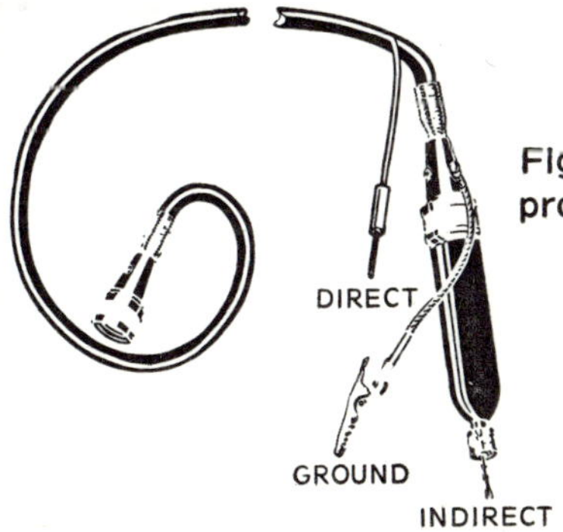

Fig. 6-5. Direct and resistive probes.

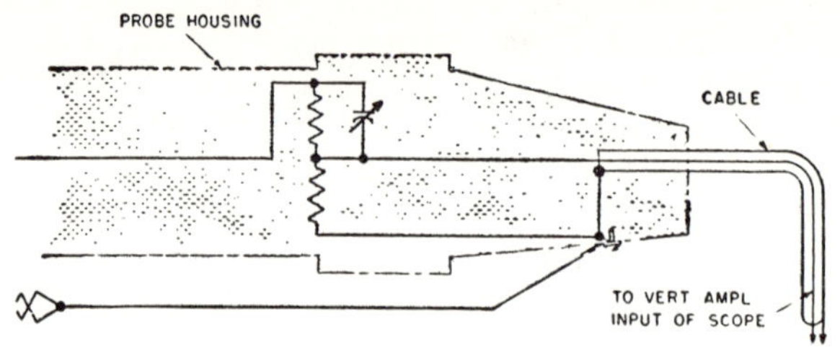

Fig. 6-6. Low-capacitance probe.

One end of the test probe is terminated with plugs that connect to the scope jacks. The inner metallic sheath of the coaxial lead is tied to the ground connector. At the probe end, this sheath will be connected to the chassis ground by means of an alligator clip, while the center conductor is touched to the test point.

A variation of the direct probe is the **resistive isolating probe**, which is essentially a direct probe with a series resistor. It is used to suppress oscillations due to feedback when the probe is touched to an active circuit.

Low-Capacity Probe

The low-capacity and high-impedance probe shown in Fig. 6-6 is useful for signal tracing in high-impedance, high-frequency, and wideband circuits, without overloading the circuit under test and causing distortion. The variable capacitor allows the probe to be tuned to eliminate the effects of frequency discrimination. This probe is especially valuable when working in television video, sync, and sweep circuits.

Demodulator Probe

The demodulator probe, shown in Fig. 6-7, demodulates AM signals, allowing signal observation in i-f and rf circuits. From the illustration it can be seen that this is a basic AM detector circuit. After detection, the signal present in the AM waveform is fed to the scope, while the high-frequency carrier (or i-f) is shunted through the filter capacitor.

OPERATION

In order to better understand the operation of the wideband oscilloscope, let's look at some of the basic operating techniques and actual applications. Although the description given here is for a particular scope, the EICO 460, you may wish to duplicate these test setups to become better acquainted with your own equipment. Later we will discuss some troubleshooting techniques using the scope. Here we are concerned more with the basics of scope operation.

Scope Setup

Without a doubt, waveform observation is the most frequently used application of the scope. But the EICO 460 can not only be used to make general observations as to waveshape, frequency, and timing, it can also be used for peak-to-peak voltage measurements.

To set up the EICO 460 for operation, the **intensity, vertical gain,** and **horizontal gain** controls are set at the full counterclockwise position (Fig. 6-4). The **focus, vertical,** and **horizontal position** controls should be set at the center of their ranges. (The position of the remaining controls is unimportant at this time.)

Insert the power cord into a 105V-125V ac outlet. Turn the **power** control (located on the scale light knob) clockwise, and allow the scope to warm up for about a minute. Gradually turn the **intensity** control clockwise until a spot appears somewhere on the screen of the crt. Should the spot fail to appear, readjust the **vert pos** and **hor pos** controls slightly. Adjust these same controls until the spot is centered in the graticule; then adjust the **focus** control for the sharpest image. For every setting of the **intensity** control, there is a best setting of the **focus** control. The finest focus is usually obtained at low intensity.

Set the **hor selector** at the **+ sync sweep** position and advance the setting of the **hor gain** gradually. Note that the spot extends to a horizontal line; this is the linear horizontal sweep. Set the **hor selector** to **ext hor**. The horizontal line returns to a spot, as the horizontal amplifier is now connected to the **hor input** terminal. Any signal applied to the **hor input** terminal will cause the line to lengthen horizontally in proportion to the peak value of the signal.

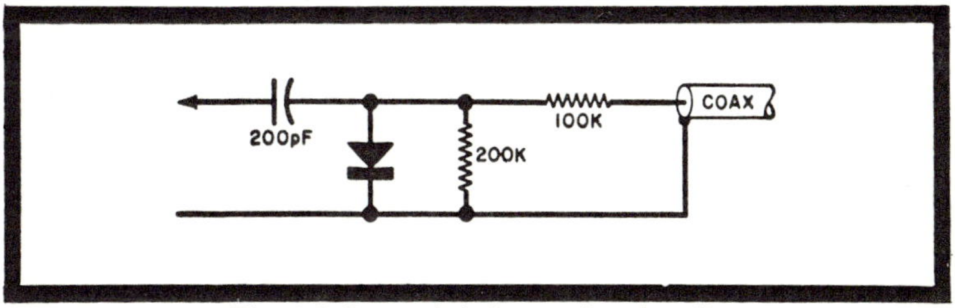

Fig. 6-7. Demodulator probe circuit.

New set the **hor selector** to the **60 Hz** position under the **hor** arc. A horizontal line representing the 60 Hz sine sweep should appear on the screen of the **crt**. The phase of this sweep is adjusted by the **phasing** control. To check this, connect a jumper between the **60 Hz** and the **vert input** terminals and adjust the **vert gain** and **hor gain** controls until the pattern is a few inches high and a few inches across. The pattern on the screen should be a stationary ellipse, because the vertical and horizontal inputs have identical frequencies. Rotation of the **phasing** control should cause the ellipse to broaden or flatten out.

Return the **hor selector** to the **+ sync** position, set the **vert attenuator** at **100**, the **sweep freq** control between **10** and **100**, and the **sweep vernier** to **10**. Leave the jumper connected between the 60 Hz and **vert input** terminals. Now adjust the **vert gain** and **hor gain** controls until the pattern extends about two-thirds of the width and height of the screen. The pattern will not be clear, because of its rapid horizontal drift. Advance the setting of the **sweep vernier** control gradually until a single, stationary sine wave is on the screen.

In rotating the **sweep vernier**, notice that the drift of the pattern slows down as certain critical frequencies are approached, and then reverses when the critical frequency is passed. At these critical frequencies, a clear pattern can be obtained. These sweep frequencies are submultiples of the signal frequency. The pattern may be locked in at submultiples of the signal frequency when it is desired to view more than one cycle of the signal. The sweep frequency is equal to the signal frequency divided by the number of complete cycles displayed. For example, if two complete cycles of the 60 Hz signal are displayed, the sweep frequency is 30 Hz. At low sweep frequencies, flickering of the pattern is normal, due to the slow writing speed of the spot.

Although this checkout procedure is specifically intended for use with the EICO 460, it is general enough to be used as a guide for setting up most wideband service scopes.

Waveform Observation

Once the scope has been adjusted and checked out, waveforms may be observed, using the following guidelines.

This method of scope operation is achieved by setting the scope up as just discussed. A jumper is connected between the balanced vertical input and ground, and the test probe lead is connected between the **vert input** and ground. The ground clip of the test probe is connected to the chassis ground of the circuit under test, and the probe tip is touched to the appropriate test point. For example, assume that you are trying to observe the signal at the collector of the transistor amplifier in Fig. 6-8. Set the **hor selector** to the desired sync polarity (positive or negative under the **sync sweep** arc). If the negative position is chosen, the scope sweep will sync with the negative-going portion of the waveform, while in the positive position, sync is achieved on the positive-going portion.

Set the **sweep selector** to a frequency that will display approximately two full cycles of the input waveform. The **sweep vernier** control is used for the fine setting of sweep speed. Actually, the sweep speed setting will depend on the purpose of the waveform observation. If you are interested only in the general characteristics of the waveform, several cycles may be displayed. In this case, the sweep speed may be set for some subharmonic of the observed waveform. For more detailed observations, a larger, more detailed presentation is needed; thus the sweep speed is set closer to the actual frequency of the displayed waveform. As sweep speed is increased, the total number of cycles displayed is decreased and each complete cycle occupies a larger portion of the screen. For example, in Fig. 6-9, three different presentations of the same waveform are displayed. The observed waveform is a square wave at a frequency of 1000 Hz. In display A, the sweep speed of the scope is approximately one-third that of the input waveform; therefore, three complete cycles are displayed. Example B shows the resulting display when the sweep speed is about one-half that of the input waveform; two complete cycles are displayed. As sweep speed approaches the speed of the input waveform, the display is reduced to one complete cycle, as shown in example C. In most radio and

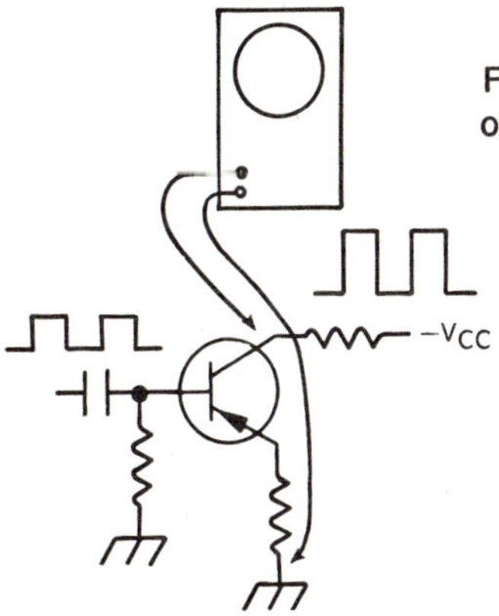

Fig. 6-8. Observing the output of an amplifier.

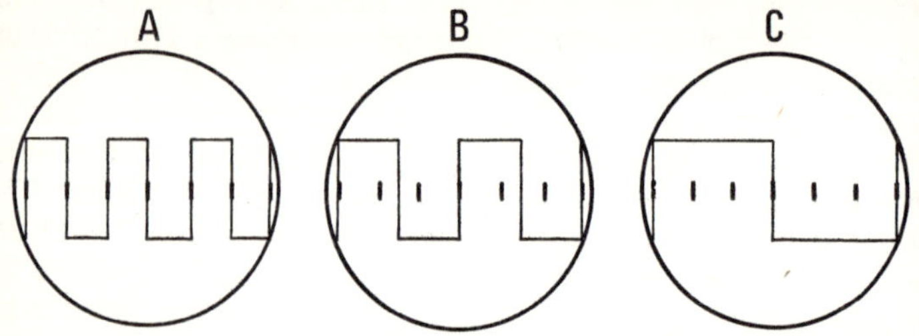

Fig. 6-9. The effects of sweep speed on the waveform display.

television service work, waveforms are displayed with a sweep speed of about one-half that of the observed signal, so that two full cycles are displayed.

Once an in-sync waveform is displayed, the vertical and horizontal settings are refined to achieve the desired size.

VOLTAGE MEASUREMENTS

Scopes such as the EICO 460 can also be used for peak-to-peak voltage measurements. After the scope has been initially set up for waveform observation, set the **vert atten** to the **cal** position. This mode of operation injects a 0.4V peak-to-peak sine wave at the power line frequency into the vertical amplifier, at a point following the vertical step attenuator but preceding the **vertical gain** control. Now set the **hor selector** to the positive position and the **hor gain** to zero. A vertical line should appear on the crt. Adjust the **vert gain** and **vert pos** controls until the vertical line extends exactly four major divisions on the graticule.

The sensitivity of the vertical amplifiers is now standardized, and direct peak-to-peak voltage measurements can be made. During the measurements, it is imperative that the **vert gain** control not be touched, as this would disturb the calibration of the vertical amplifiers. Any disturbance of the vertical gain will make it necessary to recalibrate the vertical amplifier. The **hor gain** control will have to be readjusted to display the waveform once a signal is applied to the vertical input terminals.

With the vertical amplifiers calibrated in this manner, peak-to-peak voltage readings can now be taken by setting the **vert atten** selector to a position that gives a pattern under four

Sverdrup

From the Desk of . . .

SHANNON BYRD

major graticule divisions in height. Use the **vert pos** control to position the bottom of the pattern on the lowest graduations of the graticule, as shown in Fig. 6-10. If the **vert atten** selector is set to the **1** position, full-scale deflection (four major graduations) means that the peak-to-peak voltage is 0.4V. At the **10** position, full-scale deflection is 4V; and at the **100** position, it is 40V. At the **1000** position, full-scale deflection is equal to 400V.

The peak-to-peak voltage is computed by dividing the actual vertical deflection by the full-scale deflection and multiplying by the value of full-scale deflection. For example, the waveform in Fig. 6-10 occupies 2.5 divisions. If the **vertical attenuator** is set to **100**, full-scale deflection equals 40V; therefore, the waveform has a peak-to-peak value of 25V, because

$$2.5 / 4 \times 40 = 25$$

It is possible, by using a slightly different calibration procedure, to measure peak-to-peak voltages in excess of 400V. Instead of adjusting the **vert gain** control for a deflection of four graduations, use a convenient fraction of full scale. This deflection should be as many times smaller than full scale as the expected value of the unknown voltage is larger than 400V peak-to-peak. The **vert atten** control is then set at 1000, and the peak-to-peak deflection of the unknown voltage can be read on the graticule.

The ratio of this deflection to the deflection set at calibration is then multiplied by 400 to obtain the amplitude of the unknown voltage. For example, assume that the deflection set at calibration is two major divisions, instead of four. When the **vertical attenuator** is set at **1000** and the peak-to-peak

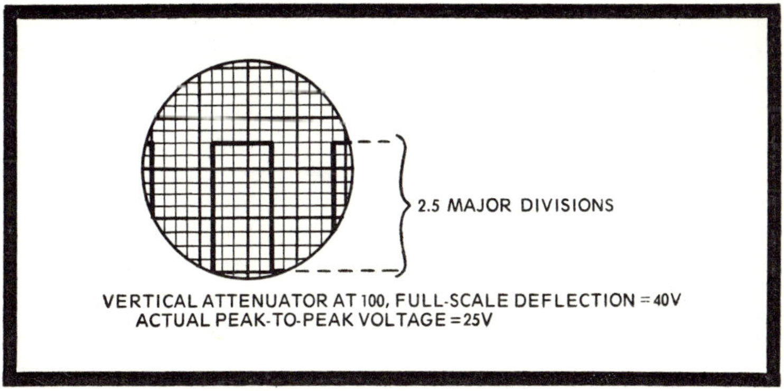

Fig. 6-10. Making peak-to-peak voltage measurements.

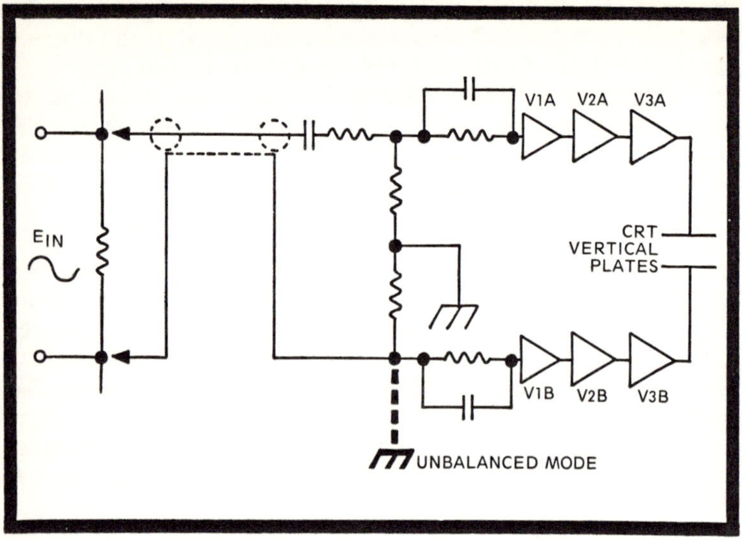

Fig. 6-11. The balanced vertical input.

deflection of the unknown signal is four major divisions, the actual voltage is

$$4/2 \times 400 = 800\text{V peak-to-peak}$$

USING A BALANCED INPUT

The use of a balanced vertical input can greatly improve scope performance in a circuit that contains a high noise level. In Fig. 6-11, we see the path of the vertical signal from the point of observation, through the probe test leads and three stages of amplification, to the vertical deflection plates of the crt. In the normal, or unbalanced, mode of operation, a clip grounds the input of amplifier VIB and thereby causes a steady-state output from V3B connected to the lower deflection plate. Any input signal is routed through the upper string of amplifiers and the upper vertical plate.

Now assume that there is some stray noise pickup in the chassis or test leads. This noise is fed through the same amplifiers as the signal, and the display will be signal plus noise. If the ground clip (heavy dashed lines in the figure) is removed, then we have a little different setup. As the top of the resistor across which the signal voltage is being taken goes positive, the bottom goes negative. The signals at the inputs of V1A and V1B are 180 degrees out of phase. Since both upper and lower channels contain an equal number of amplifiers, these signals will arrive at the vertical deflection plates 180 degrees out of phase. They will aid in deflecting the beam, since one plate will be positive and pulling the beam, while the opposite plate will be negative repelling the beam.

As the signal passes through the test leads on the way to the scope, it may pick up some noise along the way. Since there is only one component of the noise voltage, it will be applied to only one deflection plate; therefore, the influence of noise on beam position will be less than the influence of a signal input of equal amplitude. Also, any common-phase noise picked up by both test leads will be canceled, since this voltage will be applied to both vertical plates in phase. In this way, noise pickup is greatly reduced.

The balanced-input mode is obtained using the EICO 460 scope by removing the jumper between the lower **vertical input** terminal and ground (see Fig. 6-4) and setting the vertical attenuator to the **1** position under the **dc** arc. Use of the balanced input is limited by the fact that no step attenuation is possible. With balanced input, full scope sensitivity is obtained and the maximum amplitude of signal that can be observed without distortion is 1V peak-to-peak.

7 The Triggered-Sweep Oscilloscope

What is a triggered-sweep oscilloscope and what makes it different from the wideband service scopes? The sweep generator in the general-purpose oscilloscope is usually of the free-running type. A synchronizing signal injected into the time base generator at the grid can control and initiate the sweep. However, the free-running frequency of the time base generator must closely match or be some multiple of the sync signal in order to achieve a good, steady display.

While a stationary display can be achieved by adjusting the time base generator to some harmonic or subharmonic of the sync signal received from the vertical amplifier, this display may not remain stationary for any length of time. Fluctuations in either the sweep or test signal can cause the signals to fall out of sync, and if the sweep is not precisely in step with the signal being observed, the display on the crt will move. At times, it can be very difficult to obtain a stationary display, even though the sweep frequency is some subharmonic of the input signal.

Also, random pulses having low duty cycles sometimes fail to yield a stationary display. For example, let's say a pulse of 0.5 usec and 1 kHz is to be displayed. The sweep time would need to be about 1 usec in order to display the pulse, but for every pulse displayed, there are 999 equal periods of idle sweep time. This allows ample time for the time base generator to get out of sync with the observed pulse.

The triggered sweep helps overcome the difficulties of erratic sync and permits better displays of low-duty-cycle signals. This is achieved by allowing the signal under test to trigger the time base generator. During no-signal conditions, the beam is held at the left side of the screen. When a sync pulse of proper amplitude and polarity is received from the signal under test, the sweep circuit is triggered, commencing a single sweep. Following the vertical attenuator, a triggered-sweep scope employs a delay line to retard the signal being observed until the sweep has been triggered. If this delay is not used, the initial portion of the waveform will not appear on the trace, because a definite amount of time is required for the input signal to rise to a sufficient level to trigger the sweep circuits.

The B&K 1460 is a solid-state triggered-sweep oscilloscope. The triggering circuits lock the internal sweep circuit with the input signal, regardless of the frequency or complexity of the waveform, to give a good, steady display. This circuit allows the B&K 1460 to sync with as little as 10 mV of input. Although there are many triggered-sweep oscilloscopes within the price range of the small service shop, this one was selected for illustration because of my personal knowledge of it. You will find many similarities between this instrument and other triggered-sweep scopes.

BLOCK DIAGRAM DESCRIPTION

The signal to be displayed is connected to the scope through the **vertical input** jack, as shown in Fig. 7-1. From here the signal passes through the vertical attenuator to the vertical amplifier. It is then applied through the **dc bal, gain,** and **position** controls to the vertical deflection plates of the crt. The vertical amplifier is composed of direct-coupled differential amplifiers.

Part of the vertical amplifier output voltage is used as the internal triggering voltage, which is selected by the triggering circuit according to its type and polarity, and then routed to the squaring circuit. The output of this shaping circuit is differentiated to form a trigger pulse, which together with the sawtooth wave received from the holdoff multivibrator, drives the gate multivibrator circuit. The gate multivibrator, forming a dc loop with the Miller integrating circuit and the holdoff circuits, is driven from the trigger pulse to generate the sawtooth sweep voltage. This is applied through the horizontal amplifier to the horizontal plates.

The input signal may be selected as ac or dc by the **input selector** switch and applied to an 11-step attenuator. The attenuator offers an input impedance of 1M and an input capacitance of 35 pF for all positions of the **voltage/cm** switch. The output from the attenuator passes through an input protection circuit, which utilizes the diode characteristics of an FET to suppress excessive input signals to approximately ±1V.

The signal is sufficiently amplified by the vertical amplifier to be applied directly to the vertical deflection plates. Overall gain of the entire vertical amplifier circuit is approximately 60 dB, corresponding to a deflection sensitivity of 10 mV/cm.

In the phase inverter and limiter circuit, the input signal is selected as **internal, external,** or **line** by the **triggering source**

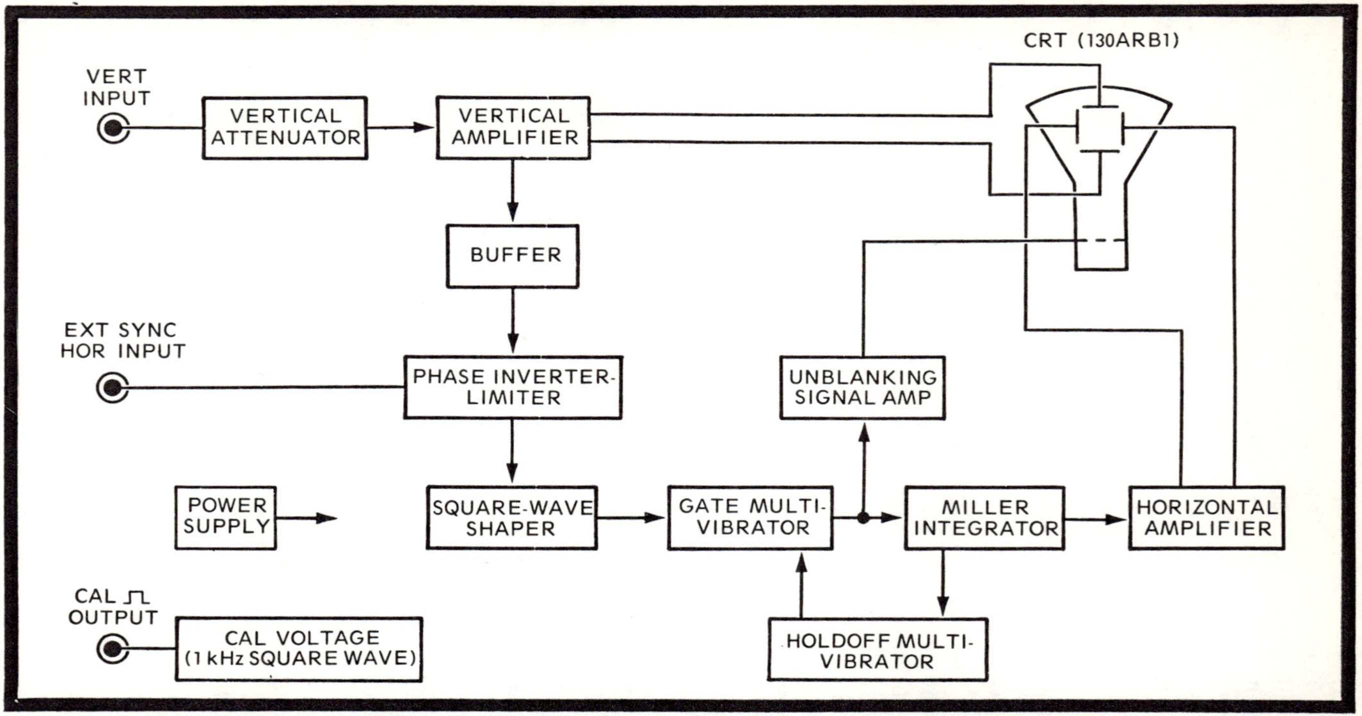

Fig. 7-1. Block diagram of the B&K 1460 oscilloscope. (Courtesy B&K Dynascan Corp.)

switch. This circuit is a sort of differential amplifier which, because of its narrow dynamic range, limits as well as inverts its input signal. The polarity of the signal can be reversed by the **triggering slope** switch. The input circuit provides an adjustment of the bias voltage by means of the **trig level** control. Thus the sync threshold level is established.

To improve the sync reliability of the B&K 1460 when viewing composite video information, a sync separator circuit is provided. This circuit is operative when the **triggering sync** switch is in the **TV-V** or **TV-H** position. Because this circuit generates output pulses corresponding to the peak amplitudes of the input signal, it is ideally suited for generating sync pulses corresponding to the tips of vertical and horizontal sync pulses of the composite video signal.

The square-wave-shaping circuit is essentially a Schmitt trigger. This circuit delivers a square wave from its output when the level of its triggering signal is raised to a certain value. The square-wave output of this circuit is coupled to a differentiator circuit, where it is transformed into the trigger signal to the gate multivibrator.

When the **trig level** control is in the **auto** position, the gate multivibrator, which normally operates as a bistable multivibrator, becomes an astable multivibrator operating at 40-50 Hz. Since this astable multivibrator oscillates weakly, it is easily pulled into synchronization with the input signal to generate the trigger signal for automatic triggering.

The gate multivibrator is a bistable multivibrator under the control of the trigger signal from the wave-shaping circuit. The Miller integrating circuit and the holdoff circuit form a dc loop which generates a sweep sawtooth voltage.

The gate multivibrator circuit is set at the threshold of operation by use of the **stability** control. As soon as the multivibrator receives a trigger signal from the wave-shaping circuit, it inverts the state of its transistors and charges the time base capacitors in the gate circuit at a rate which is determined by the combination of the time base capacitors and resistors, as selected by the **sweep time/cm** control. This charging signal is applied through the high impedance of an FET to the Miller integrating circuit. This integrating circuit not only amplifies its input, but also inverts the phase of the signal.

Part of the Miller integrator output is fed back to one side of the time base capacitors. This enables the capacitors to be charged with a constant current, thereby providing a charging waveform of good linearity.

Another part of the integrating-circuit output is coupled to the holdoff circuit. Hence, the holdoff circuit feeds back the

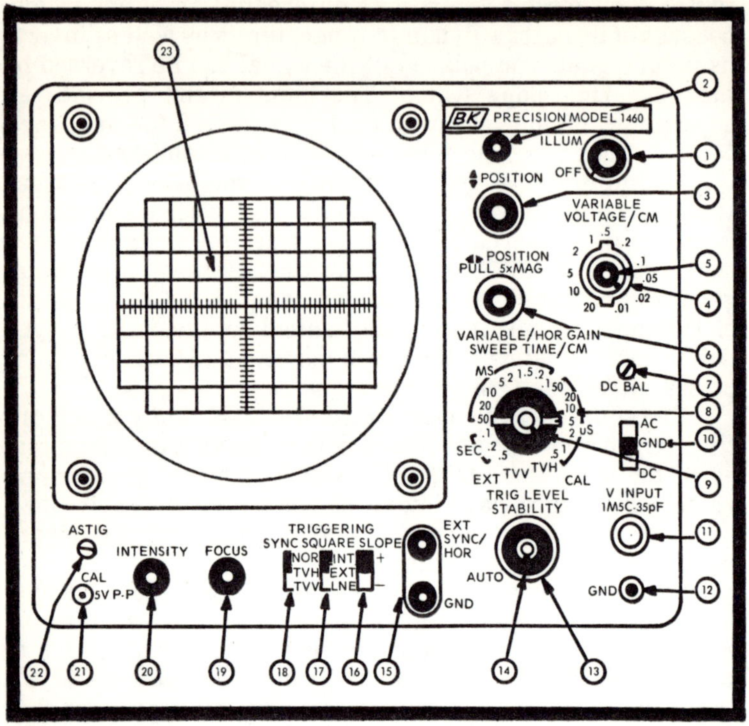

Fig. 7-2. Front panel of the B&K 1460 oscilloscope. (Courtesy B&K Dynascan Corp.)

integrating-circuit output with a delay to the input of the gate multivibrator, until the output reaches a specified level. It then inverts the state of the transistors of the gate multivibrator to their original condition to complete the sweep.

The horizontal amplifier circuit consists of an emitter follower and differential amplifier. The sawtooth output of the Miller circuit is applied to the emitter follower, the output of which is, in turn, fed to the differential amplifier.

The differential amplifier contains a variable resistor to balance the dc output voltage of the amplifier, which may be unbalanced when the **magnification** is switched from **x1** to **x5**, or vice versa. The amplifier contains another variable resistor which controls the horizontal position of the sweep. The **magnification** switch selects variable resistors which are adjusted for the horizontal amplifier gain needed for x1 and x5 multiplication.

When external sweep is used, the signal is fed from the **ext sync/hor** terminal to a variable resistor labeled "**variable / hor gain**," and on to the horizontal amplifier. The horizontal amplifier provides a deflection sensitivity of 300 mV /cm with a 3 dB bandwidth of dc to 800 kHz.

The calibration voltage circuit consists mainly of a 1 kHz multivibrator and a Schmitt trigger. The multivibrator delivers its 1 kHz output to the Schmitt trigger, which converts the signal into a square wave to provide a 5V peak-to-peak calibration voltage at the **cal** jack.

Operating Controls

The front-panel operating controls of the B&K 1460 oscilloscope are shown in Fig. 7-2. At first, this panel may seem much more complex and confusing than that of the conventional service scope; however, once you are familiar with the purpose of each control, the operation is very simple. Other triggered-sweep oscilloscopes have similar controls.

Off and Scale Illumination Control. Fully counterclockwise, this turns off the power to the oscilloscope. Clockwise rotation turns on the power and increases scale illumination.

Pilot Lamp. Lights when the oscilloscope is on.

Vertical Position Control. This adjustment controls the vertical position of the beam.

Voltage /cm Switch. This is the vertical attenuator. Coarse adjustment of the vertical sensitivity is provided in 11 calibrated steps from 0.01 to 20V /cm, when the **variable** control is set to the **cal** position.

Variable Control. Provides for fine control of the vertical attenuation between the 11 calibrated positions. In the extreme clockwise, or **cal**, position, the vertical attenuator is calibrated.

Horizontal Position Control. Rotation of this control establishes the beam's horizontal position. A push-pull switch selects x5 magnification when pulled out, or normal operation when pushed in.

Dc Bal Adjustment. This screwdriver adjustment sets vertical dc balance of the scope.

Sweep Time /cm Switch. This switch selects one of 19 calibrated horizontal sweep times from 0.5 usec /cm to 0.5 sec /cm, when the **variable-hor gain** control is set to the **cal** position. The switch also can select the proper sweep time for composite video waveforms in the **TV-H** (television horizontal) and **TV-V** (television vertical) positions. This same control will disable the internal sweep generator and display an external horizontal sweep input in the **ext** position.

Variable-Hor Gain Control. When the **sweep time/cm** switch is in the **ext** position, this control functions as a gain adjustment for the horizontal amplifiers. If the **sweep time/cm** switch is not in the **ext** position, this control acts as a fine sweep-time adjustment, controlling the horizontal sweep time between the several steps on the **sweep time/cm** switch. In the extreme clockwise position (cal), the sweep time is calibrated for the value selected at the **sweep time/cm** switch.

AC-Gnd-DC Switch. In the **ac** position, this switch blocks the dc component of the input signal. When in the **gnd** position, the input signal path is open and the input to the vertical amplifiers is grounded. This provides a zero-signal base line, the position of which can be used as a reference when making dc measurements. In the **dc** position, the vertical amplifiers are connected directly to the input signal, and the amplifiers receive both dc and ac components of the input signal.

V Input. This is the vertical input jack.

Gnd. This jack furnishes a chassis ground for connection to earth or the equipment under test.

Trig Level Control. In the fully counterclockwise (auto) position, the sweep is automatically synchronized to the average level of the waveform. At other positions, this control sets the sync level point on the waveform at which the sweep will be triggered.

Stability Control. This control sets the gate multivibrator threshold of operation to control sync stability.

Ext Sync-Hor Jack. Here, an external signal that will act as a sync pulse or horizontal sweep signal (depending on the setting of the **sweep time/cm** and **triggering-source** controls) is connected to the scope.

Triggering Slope. This switch will determine whether the sweep circuit is triggered on the negative or the positive slope of the waveform.

Triggering-Source Switch. This 3-position switch will determine from where the sweep trigger is received. In the **int** position, the waveform being observed is used as the trigger. The **ext** position connects the signal at the **ext sync-hor** jack to the trigger circuit, and the **line** position causes the sweep to be triggered by the powerline voltage.

Sync Switch. This switch provides for special sync selection when viewing television video waveforms.

Focus and Intensity Controls. These controls establish beam sharpness and brightness.

Astigmatism Adjustment. Determines the roundness of the spot on the crt.

Cal Jack. The 5V peak-to-peak 1 kHz square wave used for the calibration of vertical beam deflection is available at this jack.

Gradicule Scale. Provides calibration marks for voltage and time measurements. This particular scale is divided into 1 cm graduations. A vector overlay is used in place of this scale for vector displays.

OPERATION OF THE B&K 1460

Before the scope is used in actual circuit testing, there are a few routine adjustments that should be made during the initial starting procedure.

Before applying power, set the **off** switch to the fully counterclockwise position and connect the power cord to a 117V ac outlet. Set the **vertical** and **horizontal position** controls to their midrange positions and the **trig level** control to the **auto** position. Set the **ac-gnd-dc** switch to the **gnd** position, and turn the **off** switch clockwise to energize the scope, setting the scale illumination to the desired brightness.

After a few seconds, the crt should warm up and a trace should appear. If the trace fails to appear, turn the **intensity** control clockwise. If the trace is still not visible, turn the **stability** control fully clockwise. This places the sweep generator in a free-running mode, and no sync trigger is required to produce a sweep. Adjust the **focus** and **intensity** controls for the thinnest, sharpest trace obtainable. If the sweep is not centered, use the **vertical-** and **horizontal-centering** controls to center the sweep.

Occasionally, it will be necessary to reset the **dc bal** and **astigmatism** controls. The **dc bal** adjustment assures that sweep is not shifted when the **vertical gain**, or vertical **variable**, control is moved. Place the **ac-gnd-dc** switch in the **gnd** position and set the vertical **variable** control fully counterclockwise. Adjust the **vertical position** control to center the trace vertically on the crt. Now, turn the vertical **variable** control fully clockwise. If the trace shifts its position, return the tracc to its original position with the **dc bal** adjustment. Repeat this procedure until the trace remains stationary as the vertical **variable** control is rotated from one extreme to the other.

As mentioned previously, the astigmatism adjustment controls the roundness of the beam. If the beam is not perfectly round, the sweep trace will appear thicker in some areas than others. To adjust trace astigmatism, adjust the controls for a stable horizontal trace, setting the **intensity** and **focus** controls for the best possible trace. Now adjust the **astig** control for the finest trace.

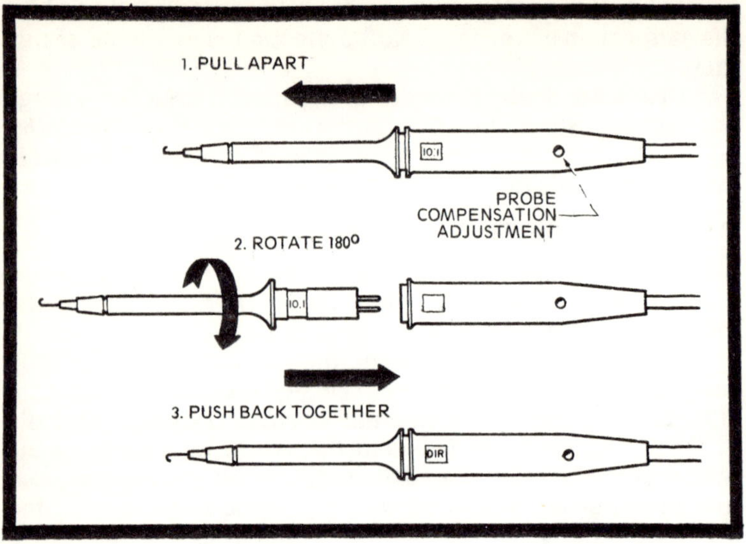

Fig. 7-3. The combination probe. (Courtesy B&K Dynascan Corp.)

The B&K 1460 comes equipped with a combination direct and low-capacity vertical input probe. For all except low-amplitude waveforms, the probe is set for low-capacity operation. In this mode, the probe has a 10M input impedance and only 18 pF of shunt capacitance. It also offers a 10:1 attenuation. The high-input impedance setting should always be used when there is a possibility of circuit overloading by the shunt effects of the scope.

For low-amplitude waveforms below 0.5V peak-to-peak, set the probe for direct operation, as shown in Fig. 7-3.

Waveform Observation

When making waveform measurements with any scope, be sure that the waveform under observation does not exceed the maximum allowable input voltage of the scope. For example, the B&K 1460 should never be connected to peak-to-peak voltages in excess of 600V.

To make waveform observations, the probe is attached to the **v input** jack and set to either the direct or 10:1 mode. Set the **ac-gnd-dc** switch to **ac** for measuring only the ac component. This will be the proper position for most measurements, and must be used if the waveform being observed includes a large dc component that might deflect the waveform off the screen. The **dc** position is used for measuring both the ac component and the dc reference; also any time a very low-frequency waveform (below 5 Hz) is to be observed.

The **gnd** position is required only when a zero-signal ground reference is required for scope calibration, such as for dc voltage measurements.

With these settings made, the ground clip of the probe is connected to the chassis ground of the equipment being tested, and the probe tip is touched to the point in the circuit where the waveform is to be measured. If the equipment under test employs a transformerless power supply, use an isolation transformer between the equipment and the ac outlet for the scope to avoid electrical shock.

This completes the first step in any waveform measurement. The exact sequence from here on will depend upon the specific conditions under which the measurements are being made. These adjustments will concern the size of the display and synchronization. Set the **voltage/cm** control to a position that gives 2-6 cm (two to six large squares on the scale) vertical deflection. The display on the crt will probably be unsynchronized at this point.

Now the sync selection switch is set to the desired type of sync. If television video waveforms are to be observed, the the **sync** switch is set to either **TV-V** or **TV-H**. For all other waveforms, the switch should be set to the **hor** position.

Another sync adjustment must also be made. If the waveform being observed is also to be used to trigger the sweep, set the **triggering-source** switch to the **int** position. (This will be the case for the majority of the waveforms observed.) When an external sync source is required, set the trigger source to **ext**. A third sync selection is offered for use with 60 Hz.

The **triggering-slope** switch is set to either the — or + position, depending on the waveform being measured. To observe composite video signals, the sweep should be synced to either the horizontal or vertical television sync pulses, depending on which specific circuit in the television receiver is being tested. Because the polarity of these pulses will vary according to the point at which they are observed, the **triggering-slope** switch may need resetting during signal-tracing operations. If the observed sync pulses are positive going, causing an upward deflection of the trace, set the switch to the negative position; if the pulses are negative going, the switch should be set to the positive position.

The **sweep time/cm** switch controls the time it takes the electron beam to sweep across the screen, thereby determining the number of complete cycles of the waveform that are displayed. Set the **sweep time/cm** control and the **variable-hor gain** control for the desired number of cycles. These controls may be set for viewing only a portion of a cycle, but the trace becomes progressively dimmer as a proportionately smaller portion is displayed, due to the increase in writing speed without a change in the sweep repetition rate. When using a very fast sweep at low repetition rates, you may wish to operate the scope with the **intensity** control set near maximum. Under these conditions, a retrace pip may appear at the extreme left of the trace. This does not in any way affect the oscilloscope operation, and may be disregarded.

The **sync** switch simplifies the sweep speed adjustment when viewing television video waveforms. The **TV-H** and **TV-V** positions automatically select the correct sweep time to display two complete horizontal lines or two complete vertical frames when the **variable-hor gain** control is set to **cal**.

If you desire to use the automatic sync features of the scope, wherein the predominant point of the waveform is automatically selected as the sync trigger, set the **trig level** control fully counterclockwise to the **auto** position. Next, turn the **stability** control counterclockwise until the trace disappears. Now turn the **stability** control clockwise just past the point where the trace reappears. This should provide a stable waveform free of jitter. Do not turn the **stability** control too far clockwise, or it will go into a free-running, nonsynchronized mode.

Should another point on the waveform be used as the sync trigger, turn the **trig level** control clockwise away from the **auto** position. Set the control by observing the waveform and note that it starts at the desired time. It may also be necessary to adjust the **stability** control. This control has three general modes when rotated from one extreme to the other. At the counterclockwise end, no trace is produced, for the sync threshold level is set so no part of the input signal is sufficient to trigger the sweep. At the clockwise extreme, the sync threshold is so low that any signal level will trigger the sweep, resulting in a free-running, unsynchronized display. The center range provides the proper threshold for synchronization and results in the desired stable presentation. The boundaries of this center range depend upon the strength and type of signal input and the setting of the **trig level** control. However, the setting of the **stability** control is not critical; it operates properly over this entire center area, and the proper setting is easy to attain.

For a closeup view of a portion of the waveform, the **horizontal position** control is pulled outward. This expands the sweep by a factor of 5 (x5 magnification) and displays only the center portion of the sweep. The **horizontal position** control can then be turned clockwise or counterclockwise to view portions of the waveform to the left or right of center.

Calibrated Voltage Measurements

The calibration accuracy of the B&K 1460 should be checked occasionally by observing the 5V peak-to-peak square wave available at the **cal** jack. This calibrated voltage source should supply exactly 5V peak-to-peak. If it doesn't, set the **voltage / cm** control to the **0.1V / cm** position and the vertical **variable** control to the **cal** position. Set the probe for 10:1 attenuation and touch the tip to the **cal** jack. Adjust the **sync** and **sweep speed** controls to display two full cycles of the square wave. The vertical amplitude of the display should be exactly 5 cm. If calibration is required, the internal **gain adj** must be reset.

Once the scope is calibrated; peak, peak-to-peak, dc, and other voltages are easily and accurately measured with the B&K 1460. After making the adjustments described for waveform measurement, set the vertical **variable** control to the **cal** position and the **voltage / cm** control for the largest vertical deflection possible without exceeding the limits of the vertical scale.

Read the amount of vertical deflection in centimeters from the scale. The **vertical position** control may be readjusted to shift the reference point for easier reading, if desired. When measuring dc voltages, adjust the **vertical position** control to a convenient reference with the **ac-gnd-dc** switch in the **gnd** position, then note the amount the trace is deflected when the switch is placed in the **dc** position.

To determine the voltage reading, multiply the vertical deflection in centimeters by the setting of the **voltage / cm** switch, as shown in Fig. 7-4. Don't forget that the voltage reading displayed on the scope is only one-tenth the actual circuit voltage when the probe is set for 10:1 attenuation. Actual circuit voltage is displayed when the probe is set for **direct** measurement.

Calibrated Time Measurement

This is the area in which the triggered-sweep oscilloscope far excels the recurring-sweep models. Pulse width, waveform periods, circuit delays, and other time durations can be easily and accurately measured with the B&K 1460. Calibrated time measurements from 5 sec down to 0.1 usec are possible. At the lower sweep speeds, the entire waveform is not visible at one time. However, the bright spot can be seen moving from left to right across the screen, which makes the beginning and ending points of the measurement easy to spot.

After setting the scope up for waveform observation, make sure the **variable-hor gain** control is set fully clockwise

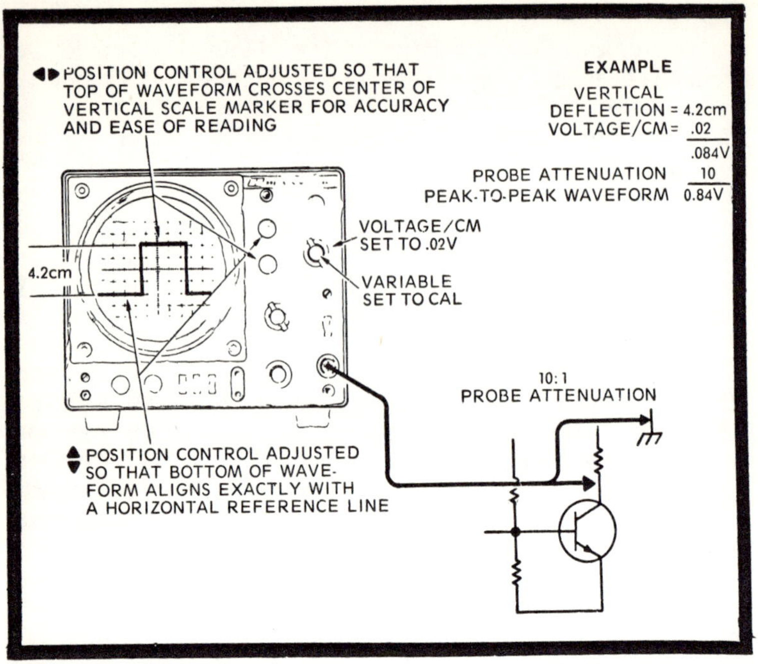

Fig. 7-4. Voltage measurement with the B&K 1460. (Courtesy B&K Dynascan Corp.)

to the **cal** position. Set the **sweep time / cm** switch for the largest possible display of the waveform segment to be measured; this will usually be one cycle. If necessary, readjust the **trig level** and **stability** controls for the most stable display. Read the amount of horizontal deflection in centimeters between the points of measurement. The **horizontal position** control may be readjusted to align one of the measurement points with a vertical scale marker as a reference for easier reading.

To calculate the time duration, multiply the horizontal deflection in centimeters by the **sweep time / cm** setting, as shown in Fig. 7-5. Just as the probe attenuation affects the vertical measurements, the **x5 magnification** switch will alter the time reading. When x5 magnification is used, the result of the previous calculation must be divided by 5 to obtain the actual time duration.

Time measurements, especially those involving delays, often require the use of an external sync pulse. In this case, the sweep is started by a sync signal from one circuit, and the waveform is measured in a subsequent circuit. This allows measurement of the delay between the sync pulse and the waveform. Set the **triggering-source** switch to the **ext** position. Using shielded cable (as short as can be conveniently used), connect the **ext sync-hor** jack to the source of the sync signal. Set the **triggering-slope** switch to the proper polarity of the sync signal, as described earlier. Readjust the **trig level** and **stability** controls, if necessary, for a stable display. Set the **sweep time/cm** control as explained previously, insuring that it is not in the **ext** position. To measure the delay between the sync pulse and the start of the waveform, determine the number of centimeters of deflection between the two, and use the same procedure as shown in Fig. 7-5.

External Horizontal Input

For some measurements, such as frequency and phase measurements using Lissajous patterns, an external deflection signal is required. In this mode of operation, the **sweep time/cm** control is set to the **ext** position and the external horizontal signal source connected through a cable to the **ext sync-hor** jack. Adjust the amount of horizontal deflection with the **variable-hor gain** control. In this mode, all sync controls are disconnected and have no effect.

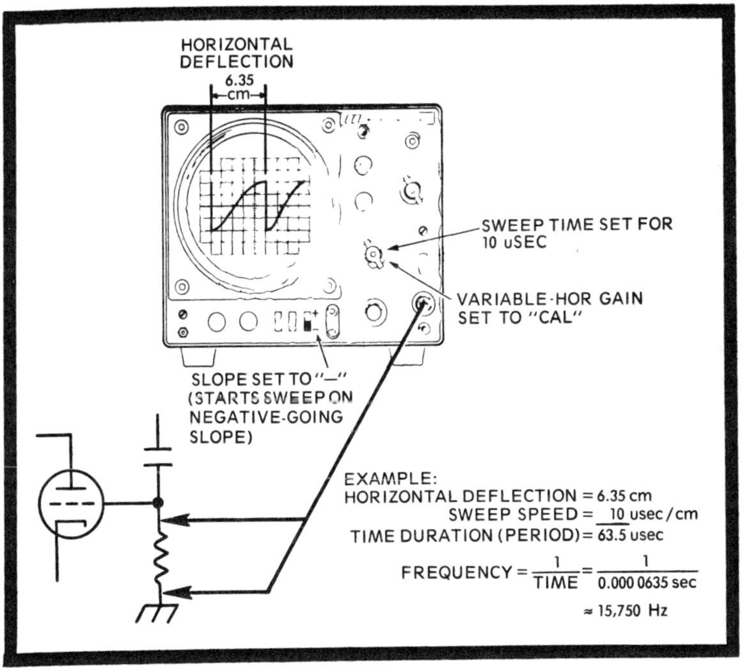

Fig. 7-5. Time measurement with the B&K 1460. (Courtesy B&K Dynascan Corp.)

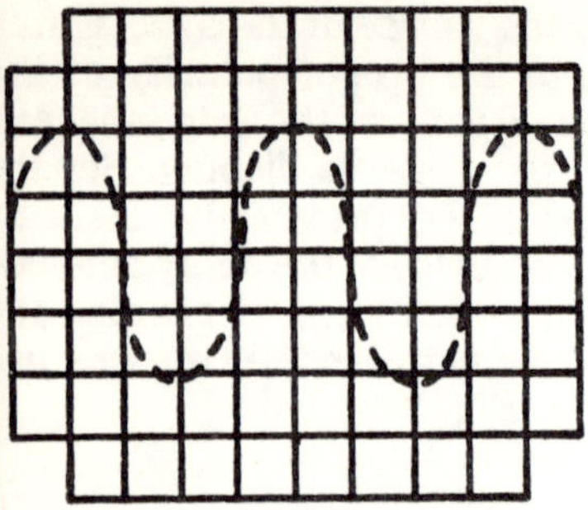

Fig. 7-6. An intensity modulated waveform using a Z-axis input. (Courtesy B&K Dynascan Corp.)

This input voltage should be about 250 mV per centimeter of deflection. Usually 2V or more peak-to-peak will provide satisfactory results.

Z-Axis Input

The trace displayed on the screen may be **intensity modulated** by a Z-axis input, to display frequency or time scale markers. A 30V peak-to-peak or greater signal applied to the **int mod** jack on the rear of the B&K 1460 will provide alternate brightness and blanking of the trace, as shown in Fig. 7-6.

Connect the vertical input probe to the circuit under test and adjust the controls for a stable display. Connect the marker source—in this case a square-wave generator—to the **int mod** jack.

DUAL-PRESENTATION OSCILLOSCOPES

Comparing two waveforms is somewhat simplified when both waveforms can be displayed simultaneously on the same scope. Using a single-trace scope, you must display first one waveform and then the other, and try to remember the characteristics of the first while observing the second. This can be confusing—and even inaccurate in cases where the waveforms are complex. Much of the confusion can be eliminated by using an oscilloscope that has facilities for dual presentations.

Dual-Beam Oscilloscope

One scope offering the capability of a simultaneous display of two separate waveforms is the dual-beam scope. The dual-beam crt can be likened to a color television crt, only with two guns instead of three. In a color television set, each beam has some controls that affect the beams independently. For example, the level of conduction of the three color guns can usually be regulated independently with the dc bias adjustments in the crt grid circuit. Similarly, the dual-beam oscilloscope has two completely separate and independent vertical channels.

A simplified block diagram of the vertical section of a dual-beam oscilloscope is shown in Fig. 7-7. This instrument utilizes a crt with **two** sets of vertical deflection plates and one set of horizontal deflection plates. Each set of vertical deflection plates is driven from its own amplifiers. Separate input paths are provided, and each input waveform is displayed independently. A bias circuit in the vertical amplifiers provides for beam separation on the crt.

Scopes of this type are expensive and not presently found in many small service shops. A more economical device that has some of the same features is the **dual-trace** scope.

Dual-Trace Scope

This scope is very similar to the dual-beam scope in function; however, there is a big difference in the actual circuitry. Two separate channels route the two input signals to the vertical output amplifier, as shown in Fig. 7-8. The vertical output amplifier and crt function just as though this were a single-trace scope, the crt having only one beam.

Each input channel is provided with its own attenuator and preamplifier. A switching multivibrator provides bias and

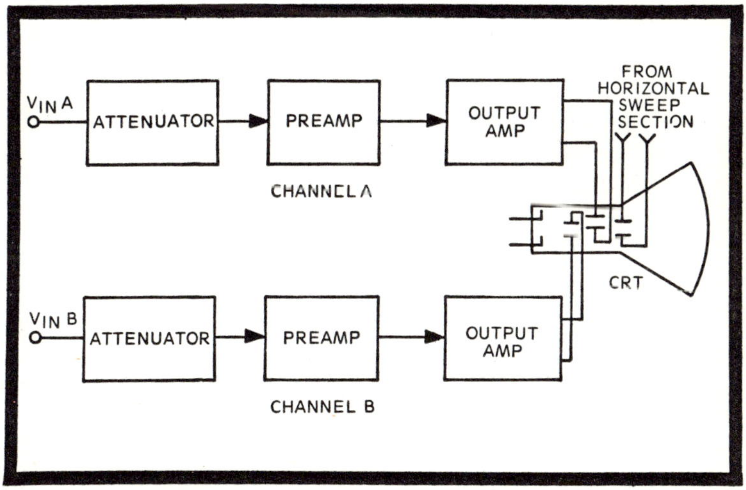

Fig. 7-7. Simplified block diagram of a dual-beam oscilloscope.

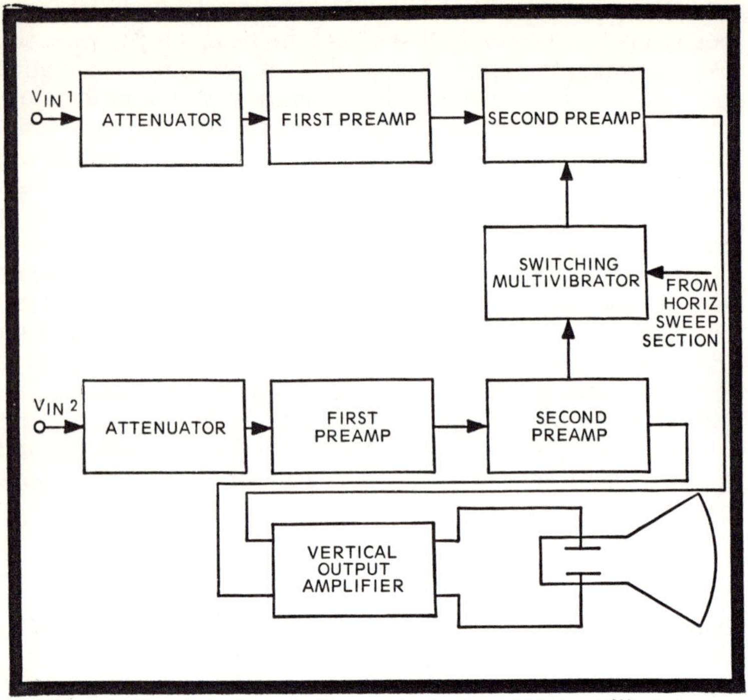

Fig. 7-8. Simplified block diagram of a dual-trace oscilloscope.

operating signals to both channels. This multivibrator is the key to the dual-trace operation. When the multivibrator's output to channel 1 is of either polarity, its output to channel 2 is of the opposite polarity. This switching action causes the output of the two channels to be active on alternate sweeps and biases the amplifiers so as to cause the traces to be displaced from each other vertically.

A trigger pulse input to the switching multivibrator from the horizontal sweep section keeps the switching action of the multivibrator in step with the horizontal sweep. The result is two independent horizontal traces capable of displaying two separate vertical inputs simultaneously.

Usually, front-panel controls are provided for the switching multivibrator which allow it to operate in one of four modes. Two modes are single-trace modes in which either channel 1 or channel 2 may be used for a conventional single-trace triggered-sweep presentation. The third mode is the **alternate** mode; in this mode, each channel is displayed on alternate sweeps. A fourth mode is the **chopped** mode. In this mode, the switching multivibrator is set to run free at a high repetition rate, independent of horizontal sweep action. The switching action is so much more rapid than the horizontal trace writing that very small segments of each trace are painted on each horizontal sweep.

The result is generally the same as in the alternate mode. Two separate traces are displayed, separated from each other vertically.

THE ELECTRONIC SWITCH

The electronic switch is a device that can be used with a conventional single-trace oscilloscope to display two separate waveforms simultaneously. The Heath ID 101, pictured in Fig. 7-9, is a good example. I have built and used one of these little devices and am very pleased with the results. It is entirely solid state and very easy to assemble, and it offers a bandwidth of dc to 5 MHz (+1.5 and —3 dB).

The complete electronic switch consists of two input networks (with frequency compensation and attenuator networks), two preamplifiers, a square-wave generator, and a switching-output circuit (Fig. 7-10). The outputs of the two preamplifiers are applied to opposite sides of the switching-output circuit. The output from the square-wave generator alternately turns each side of the switching-output circuit on and off. Thus the two input signals from the preamplifiers are alternately applied to the single output jack.

The result is one signal that is modulated on the top and another on the bottom of an internally produced square wave.

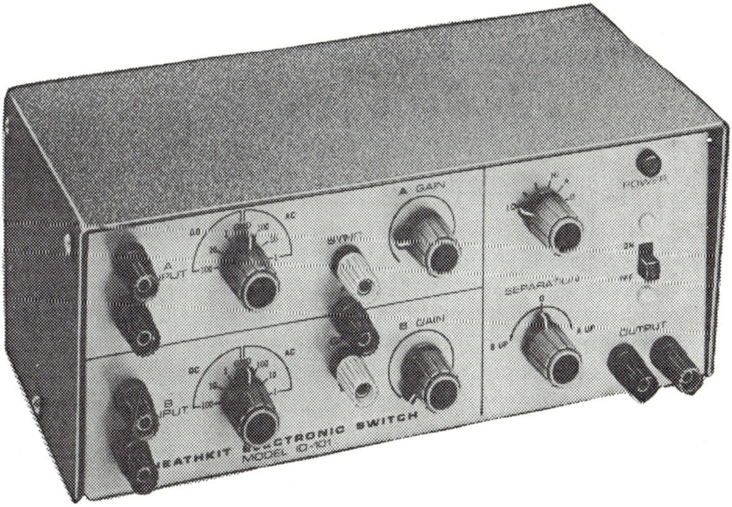

Fig. 7-9. The Heath electronic switch, model ID-101. (Courtesy Heath Co.)

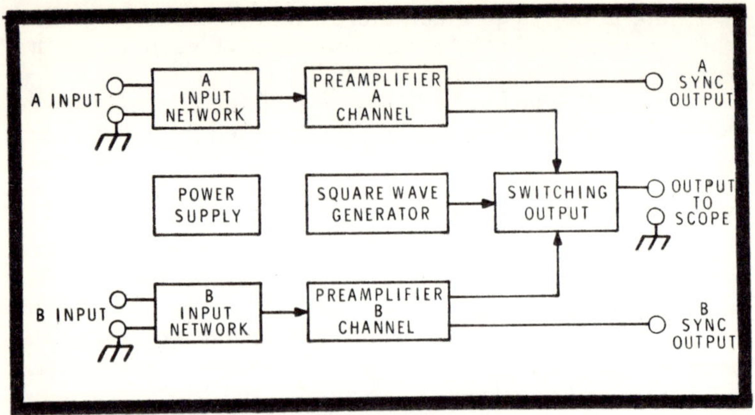

Fig. 7-10. Block diagram of the Heath electronic switch. (Courtesy Heath Co.)

This dual-modulated square wave is then coupled from the output terminals to the vertical input of an oscilloscope. When the square wave is set to the proper frequency, it does not appear itself; and only the two modulating signals are shown on the screen. The relative positions of these two signals can be changed by the **separation** control shown in Fig. 7-11, which effectively varies the amplitude of the square wave.

The **mode** switch sets the frequency of the square-wave generator. This switch is set to the position that gives the least interference with the signal displayed. The **A** and **B** positions select either the **A** or the **B** channel input signal, for a single trace on the oscilloscope.

The **input attenuator** switches adjust the input signal amplitude on both the dc and the ac ranges. The dc ranges and the dc amplifiers allow you to use the model ID 101 with a dc oscilloscope and to monitor both dc and ac components of the input signal. On the ac ranges, a capacitor is in series with the input, blocking the dc and coupling only the ac signal to the scope.

The **A** and **B gain** controls adjust the signal amplitudes within the attenuator ranges to give a desired height of the screen. The **A** or **B sync** output is used to sync the scope's horizontal sweep to the output of the electronic switch. This signal must be connected to the external sync input of the scope for dual-trace operation.

Figure 7-12 illustrates a typical application of the ID 101 and the connections to the associated equipment. In this example, the separate stereo outputs are applied to the separate channels of the ID-101 through low-capacitance

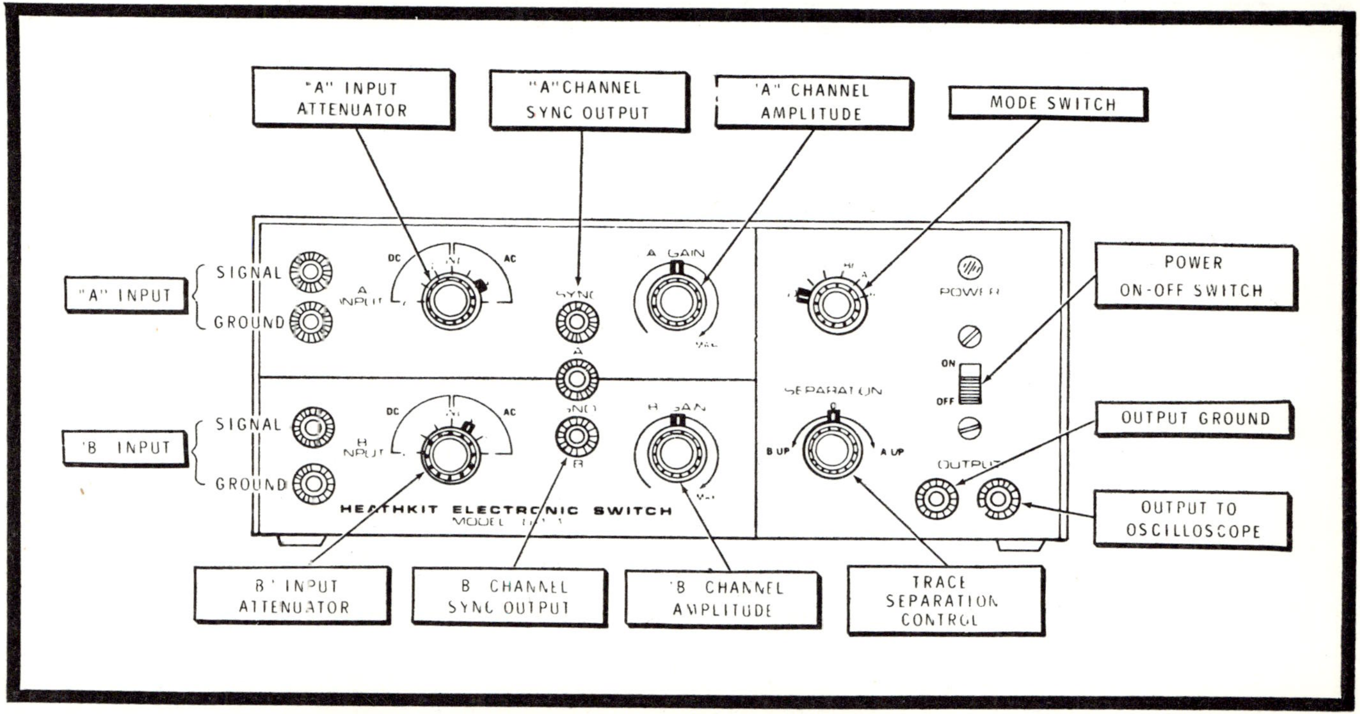

Fig. 7-11. Front panel of the Heath electronic switch. (Courtesy Heath Co.)

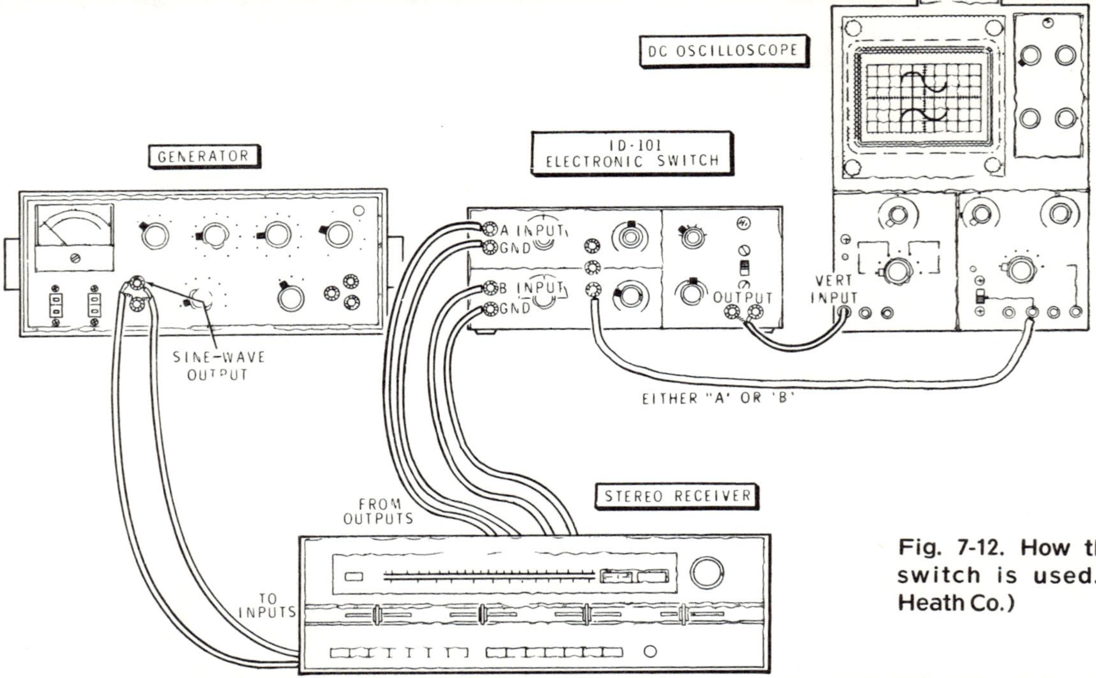

Fig. 7-12. How the electronic switch is used. (Courtesy Heath Co.)

coaxial cables. All cables should be as short as possible. The separate signals will be displayed simultaneously on the same time base.

Other applications of dual-trace operation include the following:

- Comparing an input signal to an output signal, as when checking the gain of an audio amplifier stage.
- Observing the phase relationship between two signals, such as the input and the output of a phase inverter or shifter.
- Monitoring any two time-related signals, such as the input and output signals of a frequency divider.
- Locating and comparing hum or noise between channels or between any two stages of a stereo receiver.
- Making a cross comparison between waveforms in a circuit known to be in good operating condition and the waveforms of a circuit under test.
- Comparing two dc voltage levels which vary during operation.

8 An Introduction To The Vectorscope

The vectorscope is fast finding favor with color television servicemen. This instrument is very helpful in troubleshooting the color reproduction circuits, but is somewhat limited to that area of the receiver. For that reason, many shops are equipped with oscilloscopes that can double as vectorscopes.

The reluctance of some technicians to use this relatively new instrument, in many cases, is the result of a lack of understanding of what the vectorscope does and what the display indicates. The principles of vectorscope displays are no different from those of an ordinary scope; one signal is plotted against another. Recall that in the basic scope we use the horizontal axis to represent time and the vertical axis to represent signal amplitude. With the vectorscope, the horizontal and vertical displacements are both the result of input signals; neither represents time.

The display rendered by the vectorscope is known as a **vectorgram.** Although formed in the same manner as the Lissajous patterns discussed earlier, the waveform is somewhat unusual. This is because the vectorgram is always used to compare the same two signals. That is, it is a special-application device and, unlike the other oscilloscopes, can only be used for one type of testing—color signal observation in receivers—and then only with a special signal applied.

THE SIGNAL

Figure 8-1 is a basic block diagram of a color television receiver that will be used to discuss the signals to be observed using the vectorscope. A tuned circuit in the video section removes a portion of the video signal and routes it to the input stages of the color section. This signal contains information relating to the hue and saturation of the colors to be reproduced. The brightness signal is supplied by the video section. If the receiver is processing a color telecast, the color sync signal will pass through the burst amplifier and disable the color killer. This allows the color information to pass through the color amplifiers and on to the demodulators. The

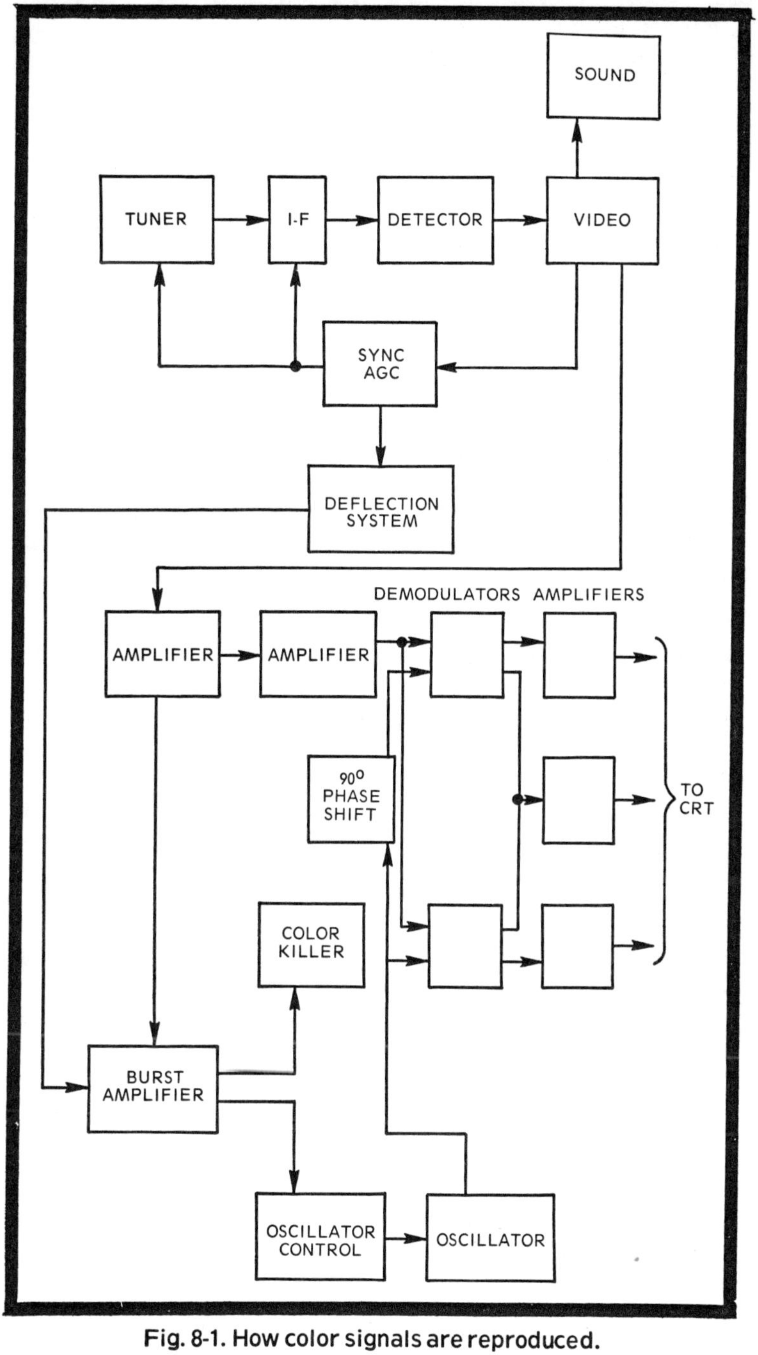

Fig. 8-1. How color signals are reproduced.

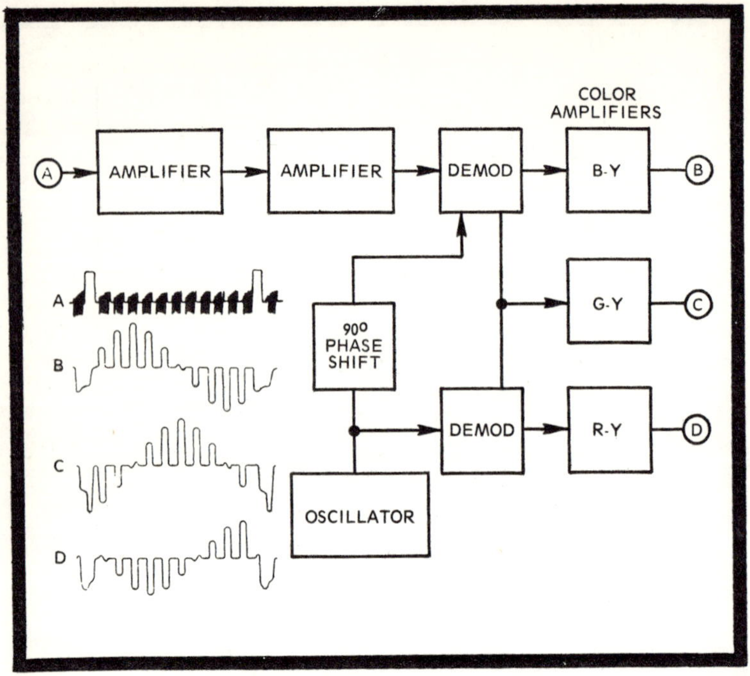

Fig. 8-2. Typical color signals.

burst amplifier also routes the color sync signal to the oscillator control circuits, keeping the local color oscillator exactly in step with the oscillator at the transmitter. In the demodulators, the incoming signal is demodulated using the locally generated signal from the oscillator. The three color signals (B-Y, R-Y, and G-Y), corresponding to the red, blue, and green characteristics of the telecast, are then formed and applied to the crt.

Since the transmitted signal is constantly changing with the scene in the telecast, it is desirable for troubleshooting purposes to use a constant signal. Such a signal for color television troubleshooting is supplied by the **color bar generator**. When this generator is connected as an input to the receiver, a predictable pattern of signal voltages results in the color section, and a definite color pattern is observed on the crt.

Figure 8-2 illustrates the overall operation of the color section. Typically, the color generator test signal results in a 10-bar color display varying from yellow-orange to green. Notice that 11 bursts are present at point A. As you may know, one pulse is used to sync the local oscillator (similar to the color sync signal in an actual transmission). The outputs of the color section are shown in waveforms B, C, and D. The amplitude of the pulses in each of these signals is of the correct value to form a distinctive color bar pattern when combined with the brightness signal in the crt. Each color bar displayed has its own particular location on the screen. The importance of that will be seen shortly.

THE DISPLAY

Now let's leave the color receiver for a moment and consider Lissajous patterns once again. Figure 8-3A is a fairly detailed illustration of the Lissajous pattern formed by sine waves of the same amplitude but 90 degrees out of phase. (For review, you may wish to refer to the earlier explanation of how this pattern is developed.) A complete cycle of both signals is present.

A complete cycle of the input voltages takes 360 electrical degrees. During this time, the electron beam is moved from position 1, in a circular manner, through 2, 3, 4, etc., and back to 1...a complete circle, or 360 degrees. You can see then, with these conditions established (two equal-amplitude sine waves, 90 degrees out of phase), the beam's position at any given

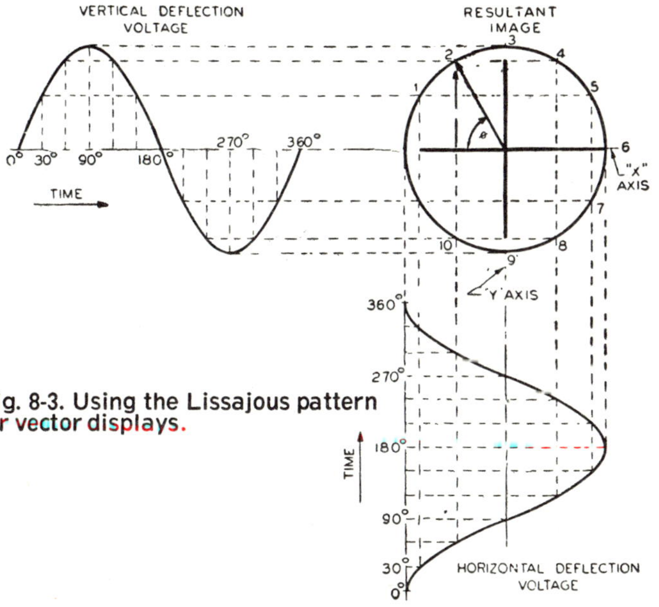

Fig. 8-3. Using the Lissajous pattern for vector displays.

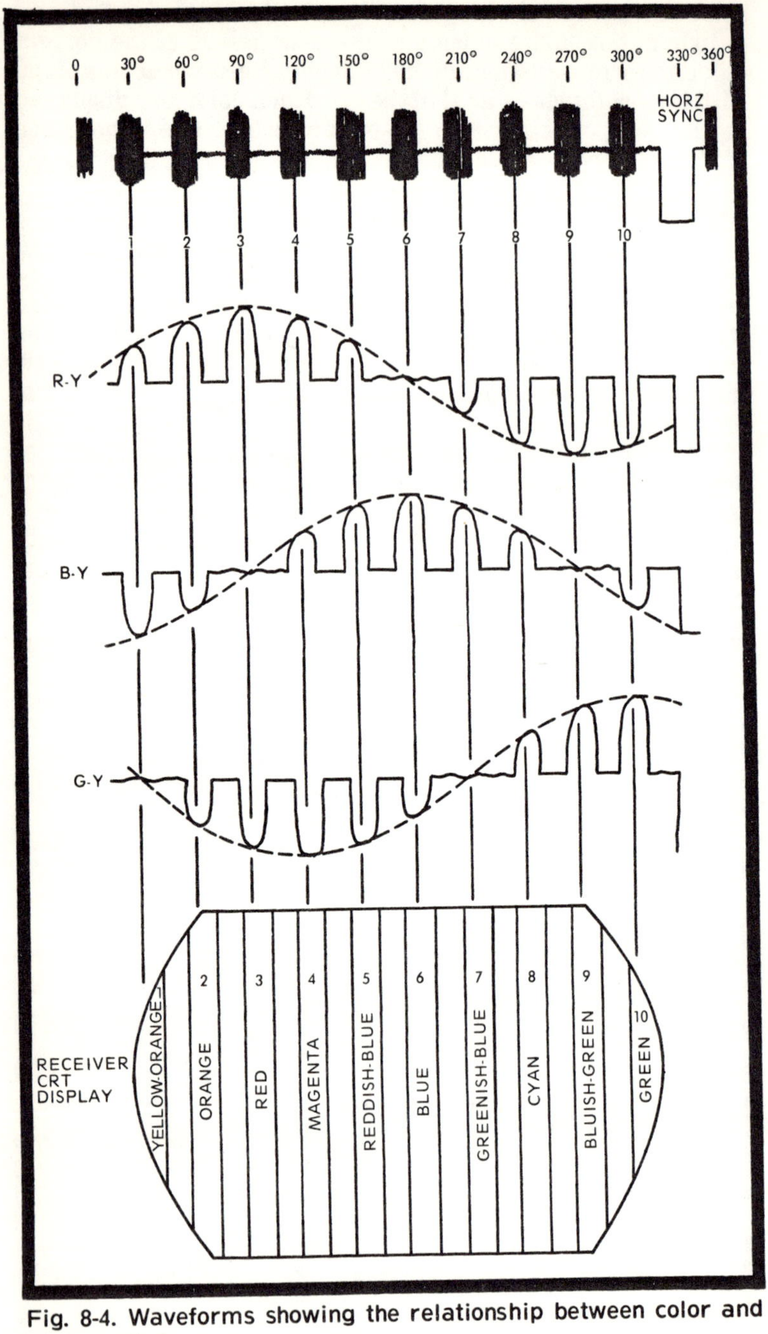

Fig. 8-4. Waveforms showing the relationship between color and phase angle.

instant represents the number of electrical degrees completed by the input signals.

Now look at Fig. 8-3B. This is an expanded version of the oscilloscope display shown in part A. Here we can see that if we draw a line extending from the center of the crt to the instantaneous beam position, we have a vector whose direction represents a specific point of time in the input cycle. As shown, the vector represents 45 degrees. With the beam at position 8, for another example, the vector will represent 330 degrees. Can't we then regard the beam as the head of a vector which is rotating in sync with the input signal? Yes, indeed.

Figure 8-4 illustrates the phase relationships among various signals in the color section of the receiver when a color bar generator is used to supply the test signals. The top waveform is the signal at the first color amplifier, while the other three waveforms represent the R-Y, B-Y, and G-Y signals (B, C, D of Fig. 8-2). The lower part of the illustration shows the pattern produced on the receiver screen with such waveforms present. Notice that for each color stripe in the display there is a specific amplitude for each of the —Y waveforms. Also, each of these pulses recurs at the same point in each cycle, one complete cycle giving one horizontal scan line on the crt. These pulses also have a specific width, which accounts for the equal spacing of the stripes.

For the time being, though, let's assume that each pulse in the three —Y waveforms is a very narrow spike and that we apply the B-Y signal to the horizontal plates of a crt, while at the same time, we apply the R-Y signal to the vertical plates (Fig. 8-5). You might have noticed that the envelopes of these signals form sine waves 90 degrees out of phase, as indicated by the broken lines in the illustration.

Starting at the 30-degree point in the cycle (pulse 1 in Fig. 8-4), we see in Fig. 8-5 that the right horizontal plate will have a sharp negative pulse, while a positive pulse is present at the upper vertical plate; this pushes the beam to the left and upward, as shown by vector 1 in Fig. 8-5. As the cycle continues, each pair of pulses at the horizontal and vertical plates causes the beam to rapidly shoot out from the center of the crt; it then returns as soon as the pulse passes. With each pair of pulses, a line (or vector) is thus formed at a specific angle on the crt. Compare the vector positions of Fig. 8-5 with those of the pulses and stripes in Fig. 8-4. It is apparent that just as each color stripe on the television display corresponds to a specific point of the waveform cycle, so does each vector displayed by the scope in Fig. 8-5.

In actual practice, the vectorgram is not a series of narrow, "pointy" vectors. The pulses have a finite width and,

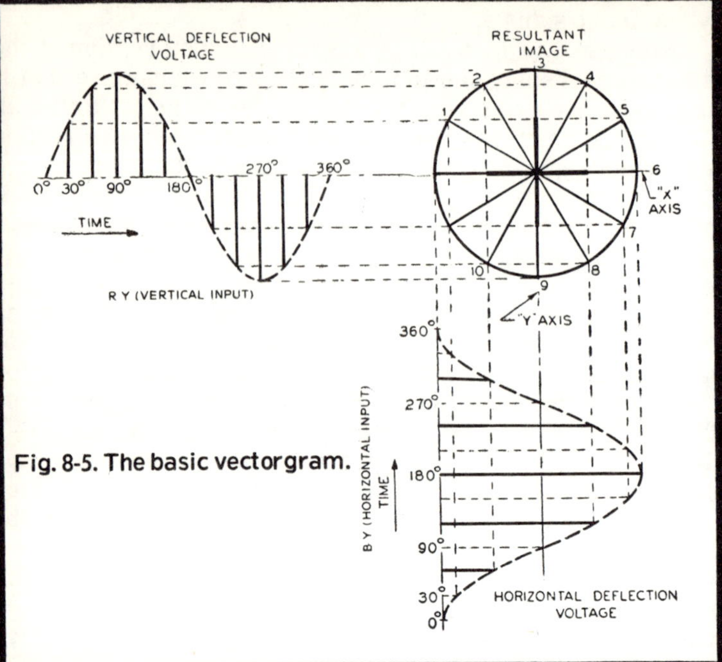

Fig. 8-5. The basic vectorgram.

therefore, the beam loops out and back instead of forming a straight line. The angle of the vectors remain the same, however, as shown in the petal display in Fig. 8-6—an ideal vectorgram. Notice that each petal represents a specific color stripe of the color bar display. No vectors occur between 330 degrees and zero, since this is the sync-burst time of the cycle.

So the vectorgram is a pattern formed by a combination of G-Y and R-Y signals that will give an indication of the color performance of a television receiver. The vector display can be studied, just as any other Lissajous pattern can, and faults in the pattern can be traced to specific stages or components. The big advantage of the vectorgram over a conventional display is the fact that you can see the overall operation of the color section in one display.

VECTORGRAM DISTORTION

All of the factors we have discussed up to this point concerning vectorscopes have been idealized. For example, the petals in the vectorgram in Fig. 8-6 would have curved sides in practice, and the tops would be rounded off. Vectorgram deviations can be attributed more to the characteristics of the signals than the characteristics of the vectorscope itself. In fact, you will find that the vectorgram will vary from receiver to receiver. Small variations should be of no concern—only those in which the petals are way out of proportion.

Before the vectorscope is connected, proper adjustment of the television controls is necessary. Assuming that the receiver is operable, tune in an active channel and adjust the chroma and tint controls for the best picture, using fleshtones

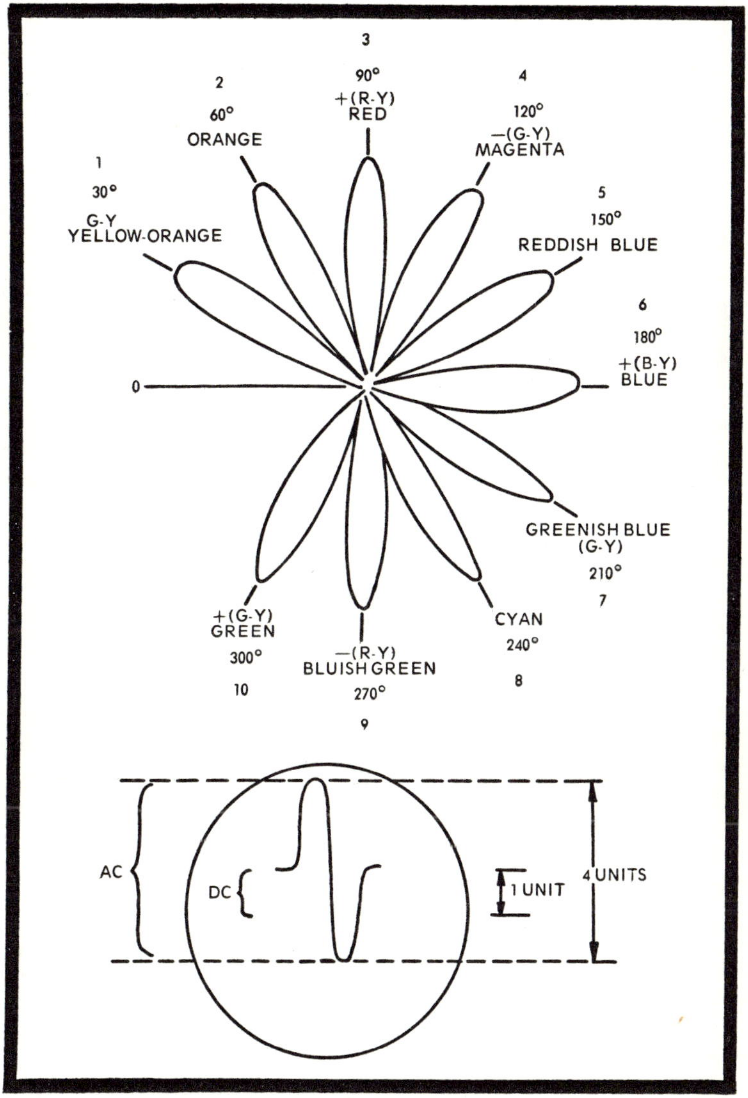

Fig. 8-6. Vectorgram showing phase angles.

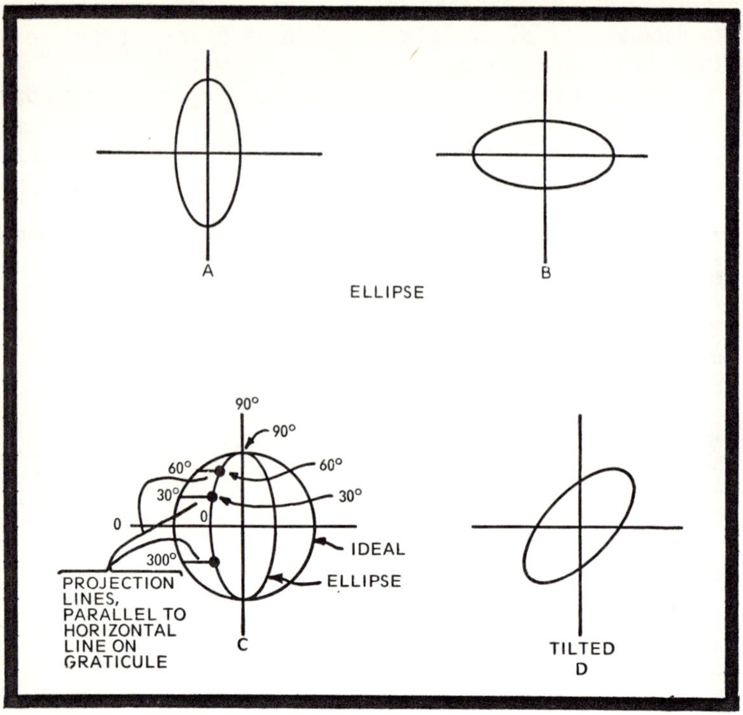

Fig. 8-7. Distortion in vectorgrams.

as a guide. Now shift the channel selector to an inactive channel, and with a color bar generator connected to the tuner input, tune in the color bar display. Tie in the vertical input to the vectorscope to the B-Y channel, and the horizontal input to the R-Y channel. The exact point in the receiver where these connections are made will vary from set to set, and it will depend somewhat on your particular vectorscope. Some service scopes are made to be used as vectorscopes by providing direct inputs from the **vertical** and **horizontal** jacks to the deflection plates. In these cases, the signal must be of sufficient amplitude to drive the plates, since the scope offers no signal amplification.

Figure 8-7 illustrates some types of distortion commonly encountered in the vectorscope method. The ellipses would be obtained by connecting the tips of the vectorgram plates. The examples in A and B occur frequently when using combination scopes as vectorscopes. When using these scopes, there is no way to regulate the gain of either channel within the test instrument and, due to the necessary differences in drive levels, the B-Y and R-Y signals may not be of the same amplitude.

The display will then be an ellipse, due to signal amplitude differences. With experience, this should present no major obstacle to proper interpretation. The tilted elliptical pattern in C is caused by a phase difference in the B-Y and R-Y signals. Part D shows how the radial degree marks can be projected horizontally onto the elliptical pattern in order to judge position.

Most vectorscope graticules are marked to indicate the correct position of the R-Y (90-degree) vector. Should this vector be slightly out of place, it can usually be brought into position by repositioning the receiver's **hue** control. In any case, refer to the manufacturer's instructions for vectorscope setup.

VECTORING WITH THE B&K 1460

Setting up the B&K 1460 scope for vectorscope displays is a snap. The first step is to equalize the gain of the horizontal and vertical amplifiers (to eliminate distortion, remember). Replace the linear graticule over the crt with the vector overlay. Connect a color bar generator to the receiver, and tune in the color bar display.

Set the **sweep time/cm** control to the **ext** position. Connect the probe cable to the **vert input** jack and another coaxial test lead to the **ext sync-hor** jack. Connect both probes to the driven element of the red gun (R-Y), as shown in Fig. 8-8.

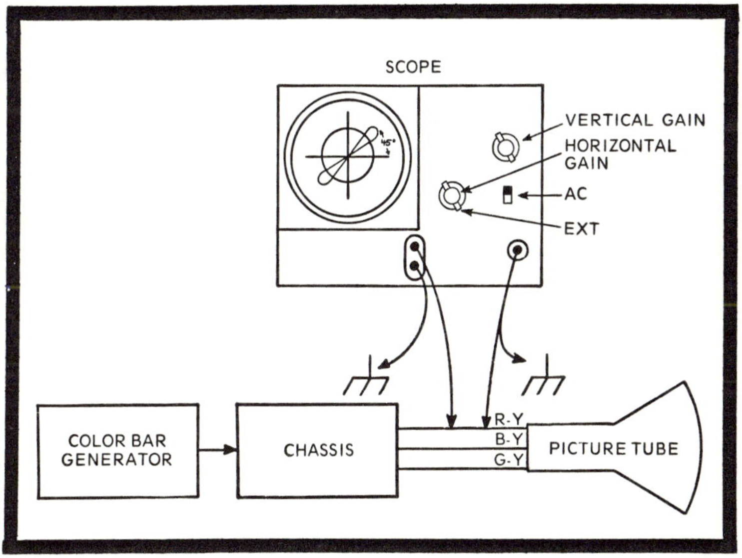

Fig. 8-8. Equalizing vertical and horizontal channels in the B&K 1460. (Courtesy B&K Dynascan Corp.)

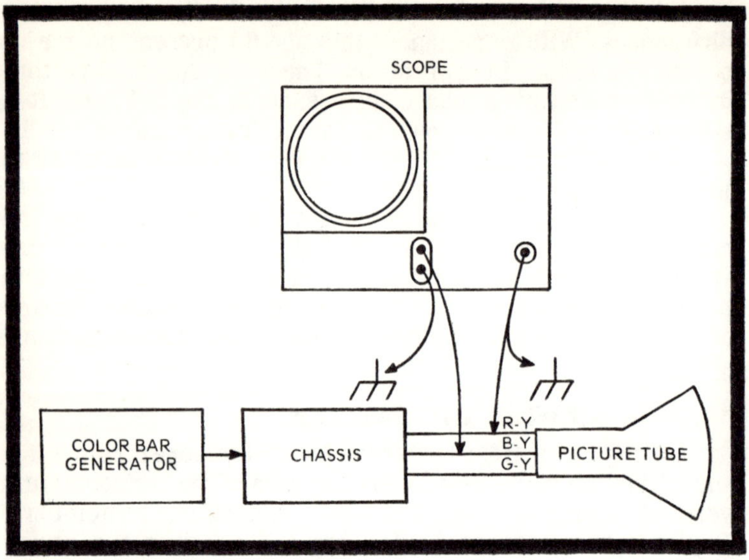

Fig. 8-9. Using the B&K 1460 as a vectorscope. (Courtesy B&K Dynascan Corp.)

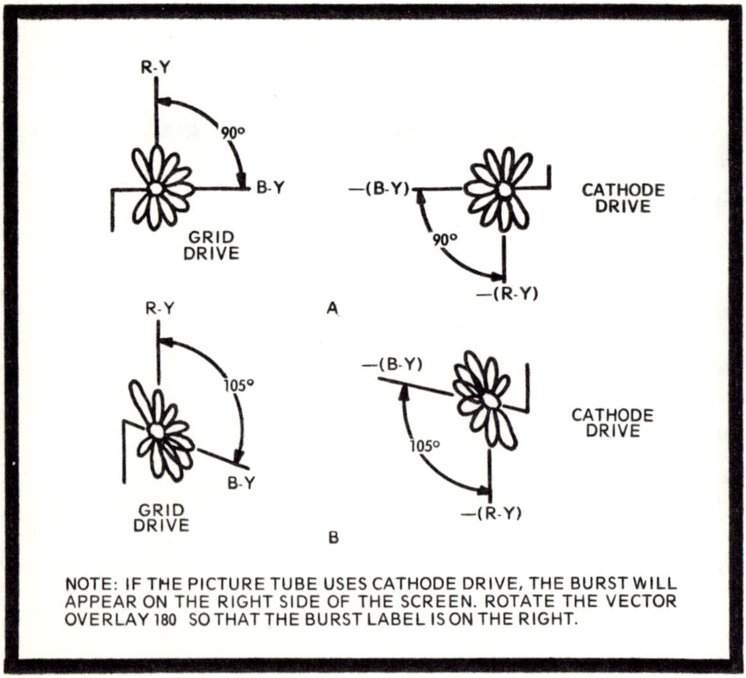

Fig. 8-10. Vectorgrams on the B&K 1460. (Courtesy B&K Dynascan Corp.)

Adjust the **vertical** and **horizontal gain** controls to obtain a compressed 45-degree pattern that approximately fills the vector graticule. Now move the horizontal lead to the driven element of the blue gun (B-Y), as shown in Fig. 8-9, and you're in business.

The vector display provides a very quick readout of the functions of the demodulators in the color receiver. However, in order to be proficient in analyzing these displays, you should familiarize yourself with the effects of the color controls on the pattern. For example, the **chroma** (intensity) control varies the size of the petals but not their position. On the other hand, the **hue** (tint) control changes the position of the petals but not their amplitude. Also, sets employing a 105-degree demodulator will produce a more elliptical pattern than the 90-degree sets. Figure 8-10 shows typical displays obtained for both sets with either grid or cathode drive.

The vector display can also be used to check the range of the receiver's hue control. It should be possible to rotate the R-Y petal. Near the center of the hue control range, the R-Y petal should be vertical. If not, a slight touchup of the color oscillator transformer may be in order. Do not attempt to make any adjustments on the bandpass amplifiers. These amplifiers are aligned using a sweep generator, and cannot usually be adjusted using the vector display.

If the set has adjustable demodulators, the vector display can also be used for demodulator alignment. Follow the manufacturer's alignment procedure, and make the necessary adjustment to obtain the correct angle between the R-Y and B-Y petals.

9 Scope Applications

Now that you have become acquainted with the operating features of your scope, it is time to look at some basic applications. In each of these cases, we will be concerned with comparing the waveforms observed with the ones given in the service literature, to determine the overall operating characteristics of a piece of equipment. In addition, we will consider some tests of individual components and how to measure specific quantities.

SIGNAL GENERATORS

Any time an oscilloscope is employed in troubleshooting, it will be used in conjunction with some signal source. This may be a signal transmitted from another location, such as a broadcast signal; a signal inherent in the equipment under test, such as a local oscillator signal; or a signal from a signal generator.

The shape of a signal generator's output voltage is determined by the type of circuits in which it is to be used. Audio signal generators, used in oscilloscope tests of audio circuits, usually have at least two output waveshapes, a sine wave and a square wave. Adjustments of the amplitude and frequency of the output are provided. The audio signal generator is an indispensible aid to "scoping" audio circuits. Injecting an audio signal into an amplifier under test, and tracing that signal through the various stages, is also one of the most basic applications of the oscilloscope.

The radio-frequency signal generator is to rf circuits what the af signal generator is to the audio section. A typical rf generator has a sine-wave output that can be adjusted in frequency and amplitude over several frequency bands. The output signal may be either amplitude modulated or unmodulated, depending on the setup. Most generators offer either 400 or 1000 Hz tone-modulated rf outputs. These generators are used in testing and aligning rf circuits.

A variation of the common AM radio-frequency signal generator offers an FM output for testing frequency-modulated circuits. Yet another rf generator—the **FM**

multiplex generator—supplies a signal for testing and aligning FM multiplex circuits. To test color televisions, **color signal generators** are available that will furnish just about any signal necessary. For radar, computer, and industrial testing, there's the **pulse generator.** The output of this signal source has pulse-type waveshapes for trigger and sync signals.

A signal generator in common use is the **sweep frequency generator**, used for the alignment of tuned rf circuits. The output of this generator is an rf signal varying about a specified center frequency. Normally, the sweep frequency generator is used in conjunction with a **marker generator**, which supplies signals at selected discrete frequencies. Some sweep generators have marker circuits in them, eliminating the need for an extra generator.

In other words, for just about any type of waveform observation with an oscilloscope, there is an accompanying signal generator to supply the necessary test signal. Typical signal generator-oscilloscope connections are described from time to time in this book.

VOLTAGE MEASUREMENTS

Just as the EVM can be used to make quantitative measurements, so can the oscilloscope. In addition, the scope can be used to measure some values—such as frequency—that are not readily measured using an EVM.

An oscilloscope with both ac and dc vertical input provisions can be used to measure either ac or dc voltage alone, or the sum of the two. In the case of ac voltage measurements, either peak or peak-to-peak values can be determined readily.

To measure voltage with any oscilloscope, the unit's vertical deflection system must first be calibrated. Usually, a test voltage for this purpose is available on the front panel of the oscilloscope. Operating instructions for the scope will give the exact procedures to be followed. Once the unit has been calibrated and the horizontal sweep speed and sync settings made, the measurement can be made. If only ac voltage is to be measured, the vertical input is set for ac only. Then the probe shield is connected to the chassis (reference point) and the probe touched to the test point, as shown in Fig. 9-1.

In most cases, the **vertical gain** control will have to remain stationary while making voltage measurements. The height of the waveform display must be controlled by the **vertical attenuator**. The peak or peak-to-peak voltage can be calculated from the resultant display.

The calculation is made by multiplying the amount of vertical deflection times the vertical **volts-per-centimeter**

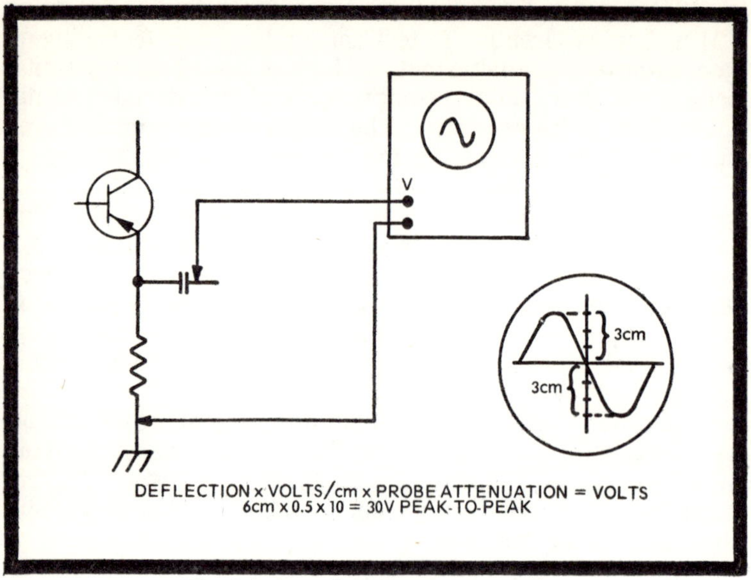

Fig. 9-1. Ac voltage measurements.

setting, times any attenuation in the probe. For example, assume in Fig. 9-1 that the waveform covers a total of 6 cm of vertical deflection, the vertical **volts-per-centimeter** switch is set to the **0.5** position, and a 10:1 probe is used. The peak-to-peak voltage is equal to

$$6 \times 0.5 \times 10 = 30V \text{ p-p}$$

Peak voltage, of course, is half the peak-to-peak value.

When only the dc voltage component is to be measured, the **vertical input** switch is set to the **dc** position. In this mode, the blocking capacitor is removed from the vertical input circuits, and both ac and dc components are measured. Notice the offsetting effect of the dc in Fig. 9-2. With the same vertical settings as above, this voltage is computed as above. The value is

$$6 \times 0.5 \times 10 = 30V \text{ p-p}$$

The peak-to-peak value is the same as before, but notice that the waveform is offset below the horizontal base line. This means that a negative dc component is present. The offset of 1 cm represents 5V; therefore, 5V dc is present in the signal.

At first, voltage measurements with an oscilloscope may seem much more difficult than with an EVM. Although some calculations are required, scope measurements have the advantage of displaying the entire nature of the voltage—frequency, shape, and both ac and dc content.

AUDIO AMPLIFIERS

Two common signals are used in audio amplifier tests using an oscilloscope. Sine-wave test signals are normally used for tests involving phase shift measurements and distortion, while square-wave tests are used for checking frequency response. First let's look at the distortion test.

In this test, apply an audio-frequency sine wave to the input of the amplifier under test. (The frequency of the test signal will depend on the frequency of the particular amplifier. Usually, the manufacturer's service literature will specify a frequency; if not, I choose one somewhere near the middle of the frequency response of the amplifier.) Connect the scope as shown in Fig. 9-3 and sync it to the test waveform. Display several complete cycles for convenient observation, but do not use such a slow sweep speed that you cannot observe each complete cycle of the test signal clearly. Do not set the amplitude of the generator output so high as to exceed the maximum input limit of the amplifier.

The purpose of this test is to reveal any differences (distortion) between the signal applied to the input and that available at the output of the amplifier. If dual-trace operation or an electronic switch is available, both waveforms can be displayed simultaneously. This makes comparison easy.

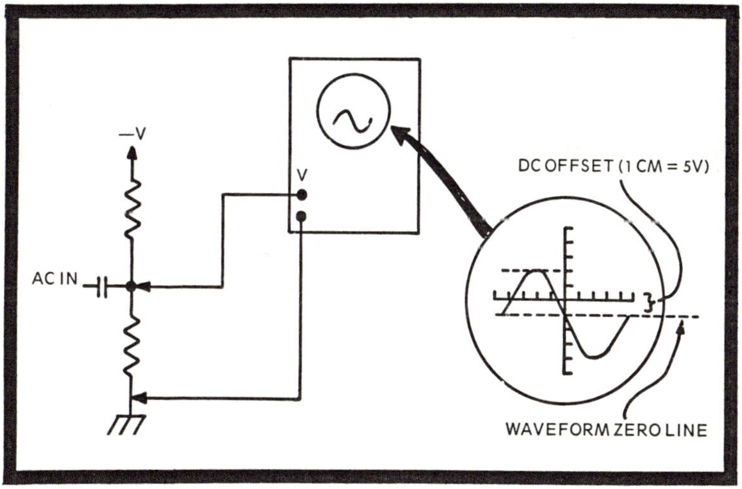

Fig. 9-2. Dc voltage measurements.

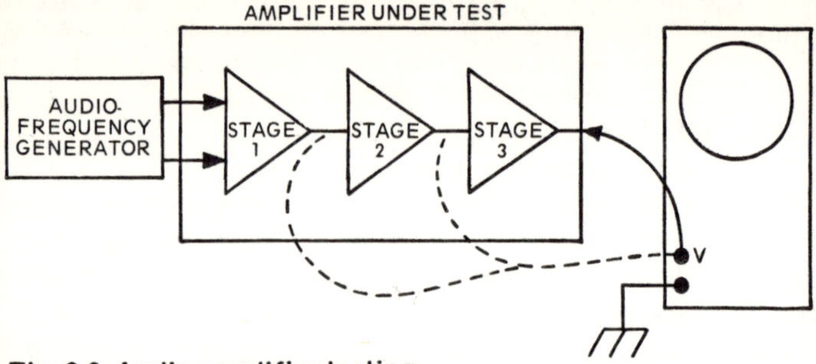

Fig. 9-3. Audio amplifier testing.

Clipping is one form of distortion that is readily revealed by this test. Positive or negative peaks, or both, are squared off.

If distortion is discovered in the output, stage-by-stage testing can be used to isolate the faulty circuit. Just have the generator connected as before, and check the input and output of each stage, starting at the amplifier output and moving one stage closer to the input each time. Keep in mind that as you move toward the generator, the signal amplitude will decrease. Do not increase the amplitude of the generator output. Instead, increase the scope's vertical gain to obtain a display of sufficient height.

With the aid of a calibrated vertical input on your scope, you can measure stage gain using the same basic setup. By comparing the input and output voltages of any one stage, you can determine if that stage is doing its job properly. To get a true picture of the gain of any amplifier, tests should be made at several frequencies—usually at about 1000 Hz intervals—over the entire amplifier bandwidth.

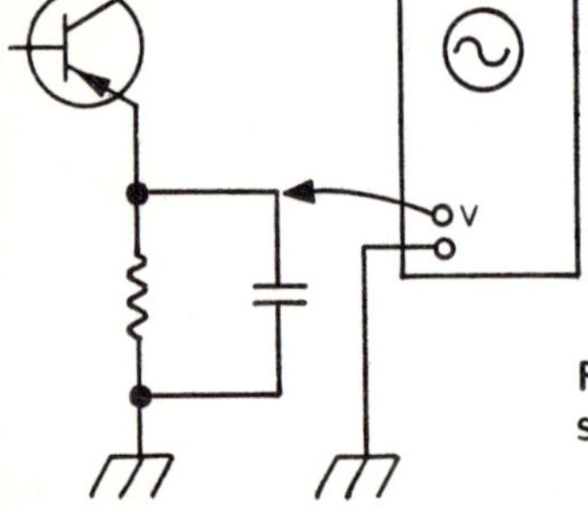

Fig. 9-4. Testing an individual stage.

Once a defective stage is isolated, it is sometimes possible to locate the faulty component using the same setup. Assume that in the situation of Fig. 9-3 we find stage No. 2 has a gain problem. A loss of gain can be caused by several conditions in a typical audio circuit. After checking the stage output and finding it low, we would move the scope probe to the emitter, as shown in Fig. 9-4. If here we see a signal of good amplitude, we have a good clue as to the problem.

The emitter resistor is bypassed by the capacitor to prevent signal degeneration. No signal should be developed at the emitter as long as the input signal is above the low-frequency cutoff of the amplifier. Evidently, then, the bypass capacitor has lain down on the job...it's probably open.

Lissajous patterns are very useful in determining the presence of phase shift in an audio-frequency amplifier. With the scope and generator connected as shown in Fig. 9-5, the amount of phase shift in the amplifier can be determined as described earlier. Notice that the scope sweep circuits are not used. Instead, the horizontal sweep is taken from the output of the amplifier, causing the output of the generator to be plotted along the vertical axis and the output of the amplifier to be plotted along the horizontal axis of the screen.

One of the easiest methods of observing frequency distortion in an audio amplifier is square-wave testing. Again, the output waveshape from the amplifier is compared to that of the signal generator. The convenience of the dual-trace presentation for this is apparent; however, effective display

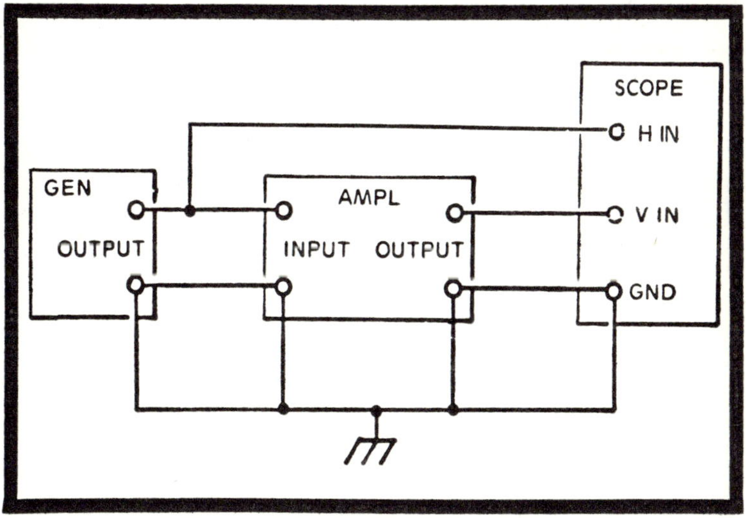

Fig. 9-5. Phase shift measurement in an audio amplifier.

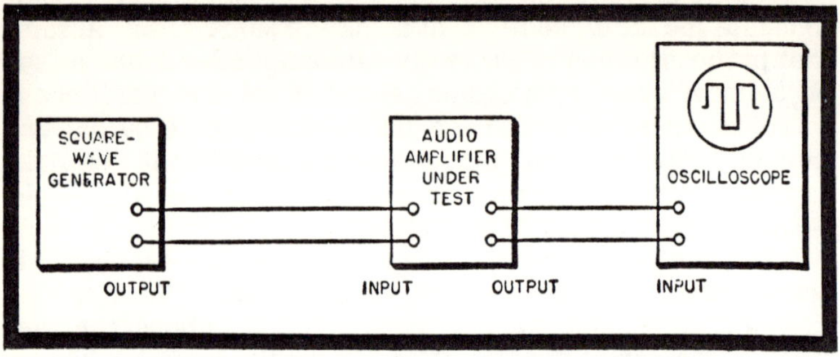

Fig. 9-6. Square-wave testing of an audio amplifier.

comparisons can be made with a single-trace scope if you keep going back to the generator output to keep in mind the test voltage waveshape.

Use a low-capacitance probe, as shown in Fig. 9-6, to prevent distortion. A switch arrangement can also be used, wherein both signals (amplifier output and generator) are brought out to a switch. Comparison is made by first displaying one waveform, then the other, by changing switch positions. Analyzing the square-wave display was discussed earlier.

CHARACTERISTIC CURVES

The active elements of electronic circuits—tubes and transistors—should be tested under actual operating conditions. Since these are dynamic devices, static tests do not

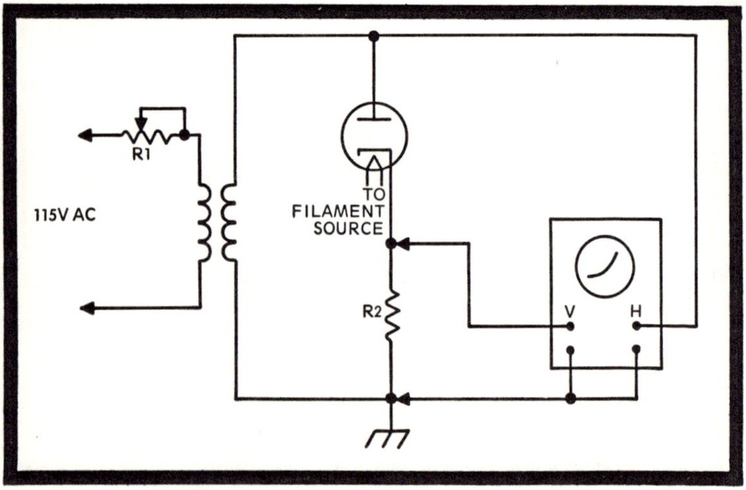

Fig. 9-7. Testing a vacuum-tube rectifier.

offer a complete picture of their condition. **Characteristic curves** are graphic displays of an active element's operating characteristics. You probably have seen characteristic curve charts in connection with explanations of tube and transistor theory. Although these curves are not often used in troubleshooting, new devices being placed on the market are making the use of characteristic curves an increasingly popular method of transistor testing.

One method of displaying the plate voltage-current characteristics of a diode is shown in Fig. 9-7. The scope is set up to use an external horizontal sweep, and the vertical input is connected across the cathode resistor. Since the current through the diode is in series with the current through the resistor, the vertical input represents the plate current of the tube. Plate voltage, supplied by the transformer, is used as the horizontal input. Then the vertical deflection of the waveform represents plate current and horizontal deflection represents plate voltage. The signal is adjusted to the proper amplitude by means of R1.

A similar test can be used with solid-state rectifiers, as shown in Fig. 9-8. The test signal is supplied across R1, and should be held within the limitations of the diode. The manufacturer's data sheet will give the specific details of each diode. Compare the results obtained in these tests with the data sheet. A typical characteristic curve for a diode is shown in Fig. 9-9. The horizontal axis represents the applied voltage, and the vertical axis the current. Forward current is shown in

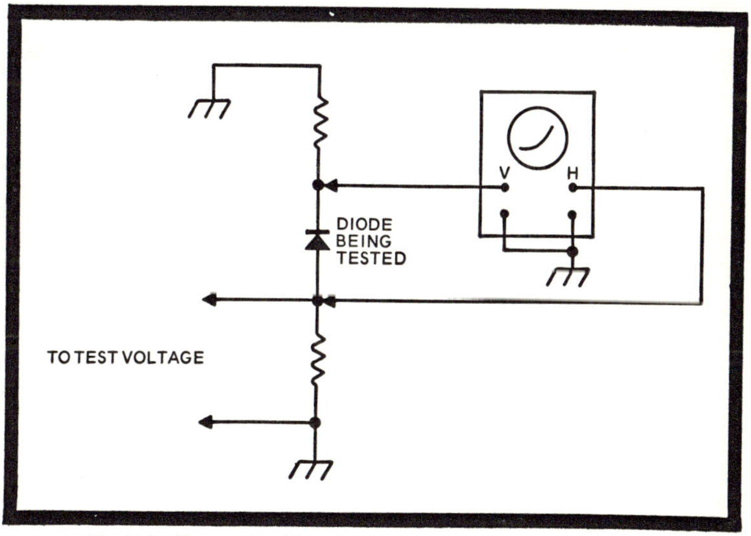

Fig. 9-8. Observing the characteristic curve of a diode.

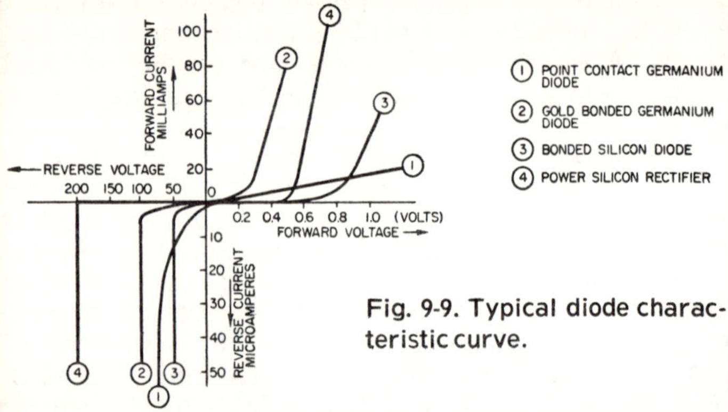

Fig. 9-9. Typical diode characteristic curve.

the upper half of the curve. Notice the zener point, where a heavy reverse current occurs. The rectifier diode may be destroyed by the back bias at this point.

The zener diode can be analyzed using the test circuit in Fig. 9-10. The resultant waveform will represent both the forward and reverse characteristics of the zener. Using this type of test, you can determine whether a zener is capable of regulating a voltage properly.

The test circuit in Fig. 9-11 can be used to show the characteristics of a transistor when used in the common-

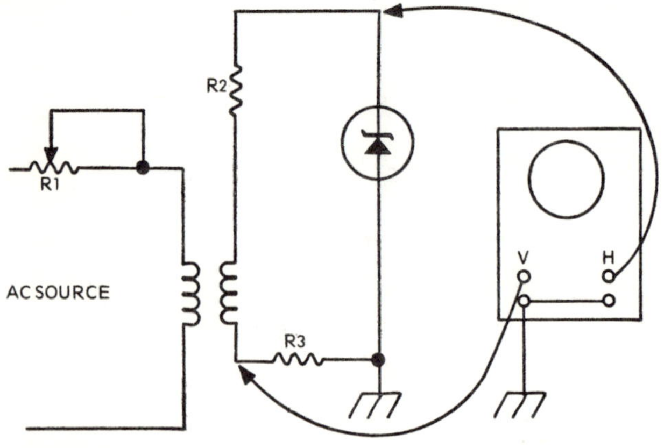

Fig. 9-10. Observing zener diode characteristic curves.

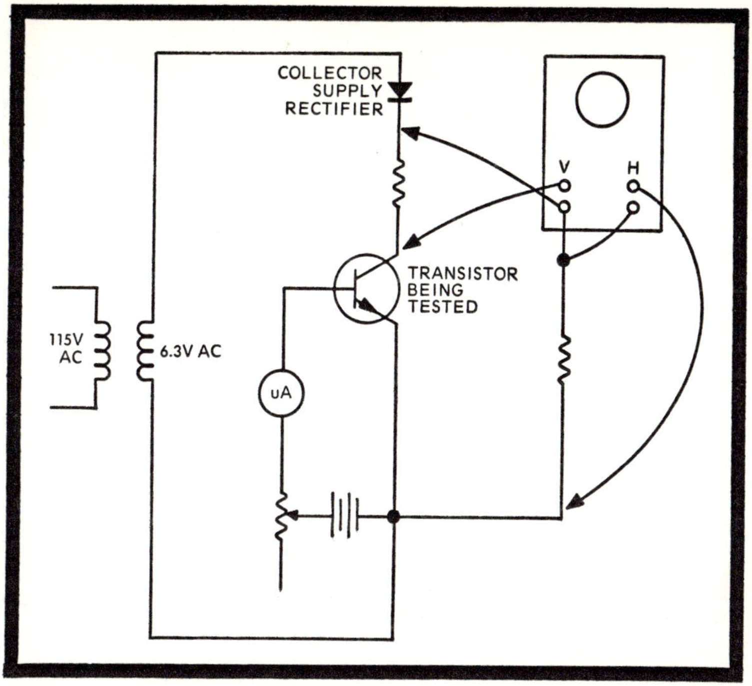

Fig. 9-11. Observing a family of curves of a transistor in the common-emitter configuration.

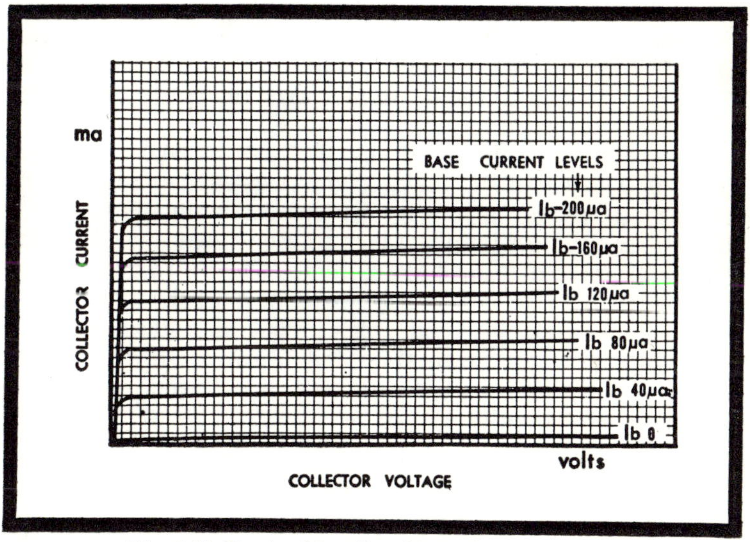

Fig. 9-12. Typical family of transistor curves.

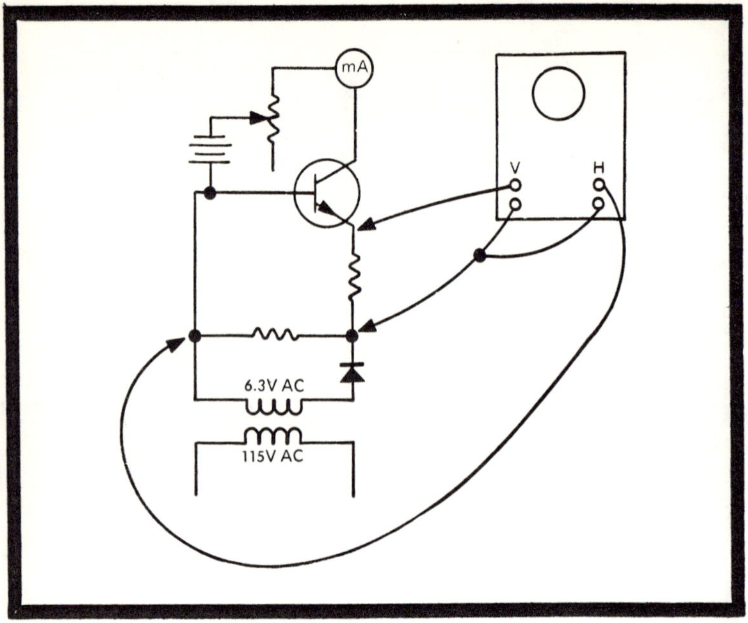

Fig. 9-13. Observing a family of curves of a transistor in the common-base configuration.

emitter configuration. The scope settings are basically the same as those used for diode testing. The vertical input to the scope is a voltage that is proportional to the collector current. A voltage representing base current is the horizontal sweep signal; therefore, the waveform will be a curve in which base current is plotted on the horizontal axis and collector current on the vertical axis.

To obtain a family of curves, the base current is changed in steps. The waveform that accompanies each change can be compared to that given in the service data. Figure 9-12 illustrates a typical family of curves. For linear operation, the transistor must be biased along the linear, or straight, portion of the curve, where a change in base current produces a proportional change in collector current. In addition to their usefulness in testing transistors, the family of curves can also be used to match transistors for use in push-pull circuits. If common-base characteristics are desired, the test setup in Fig. 9-13 applies.

MODULATION MEASUREMENT

The oscilloscope is a convenient means of observing the modulation of a carrier frequency. The Federal Com-

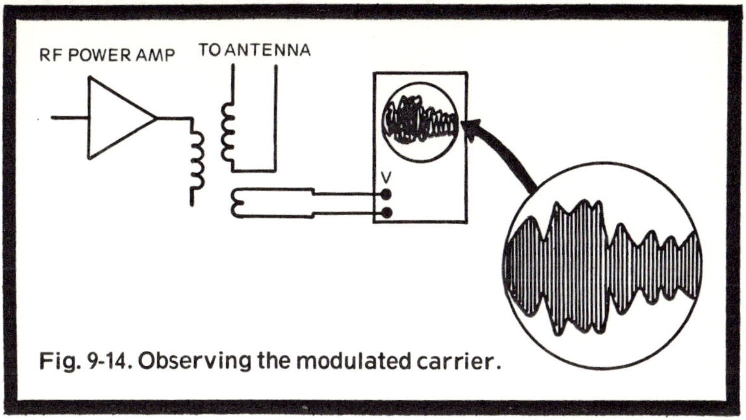

Fig. 9-14. Observing the modulated carrier.

munications Commission has formulated regulations to control the emission of electronic signals within the United States. In order to maintain equipment within the FCC specifications, the transmitter technician must have some means of observing the modulated waveform.

Due to the high voltages present in a tube-type transmitter, direct probe connections are not used. Instead, a current probe is used to obtain inductive coupling. A similar action can be seen when a common scope probe is held near the deflection yoke of a television receiver. The resultant display is the waveshape of the yoke current. In transmitter modulation measurements, the modulation enevelope is displayed (Fig. 9-14).

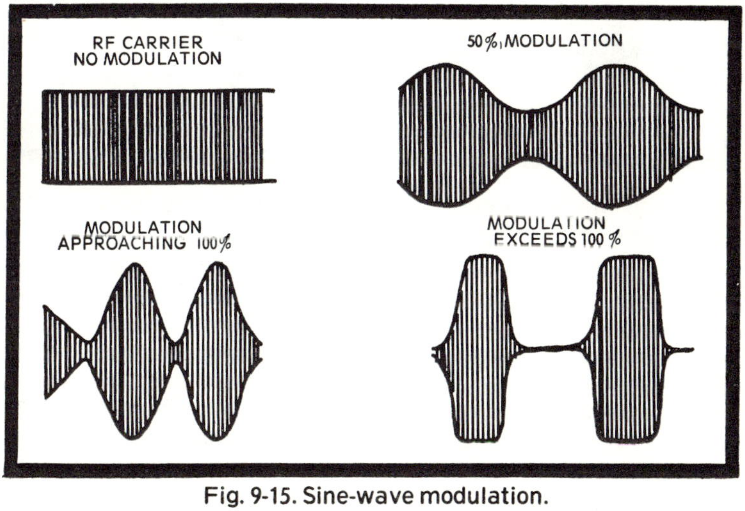

Fig. 9-15. Sine-wave modulation.

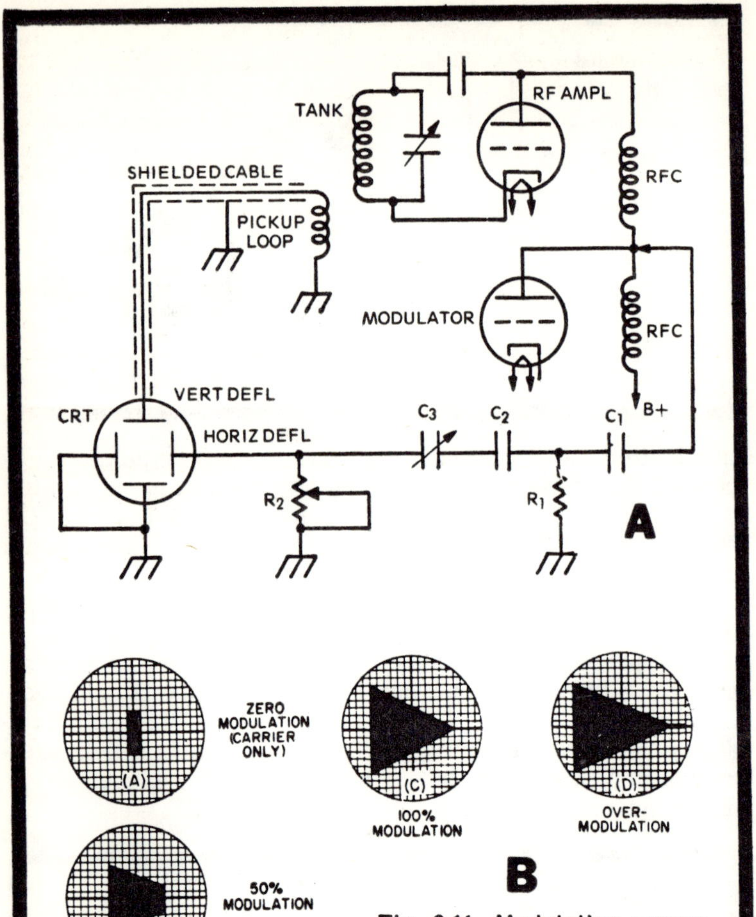

Fig. 9-16. Modulation percentage measurement.

With the equipment set up as shown in Fig. 9-14, set the **horizontal sweep, sync,** and **vertical attenuation** controls to present a suitable pattern on the crt. During normal transmissions, the modulation envelope varies at a constantly changing rate. This results in an unstable display; however, for general observation it will be satisfactory. The main indication you can observe using this method is negative **overmodulation**—a momentary loss of the modulated carrier, indicated by a bright horizontal line on the screen.

A better display may be achieved by replacing the varying modulation with a constant-frequency tone signal and syncing the scope on the modulation. Figure 9-15 depicts some waveform displays obtained in this way.

To make an accurate measurement of the percentage of modulation, the oscilloscope is connected as shown in Fig. 9-16. The **horizontal sweep** control is set to the **external** position, and the sweep signal is supplied by the modulation transformer. When the oscilloscope is connected in this manner, the display shows the modulated carrier amplitude plotted against the modulating voltage. The pattern resulting from this display is normally a trapezoid.

The longest vertical side of the pattern represents modulation peaks; the shorter side, the troughs. At 100 percent modulation, the pattern comes to a point on the trough side. Modulation over 100 percent causes the horizontal base of the wedge to be shortened and a horizontal line, or tail, to extend from the wedge, as shown in waveform D. Many broadcast stations use this method to monitor program transmissions.

With the scope connected as shown in Fig. 9-16 and modulation absent, a vertical trace (representing carrier voltage) should appear on the crt. Set the **vertical gain** control to get an arbitrary amount of vertical deflection. When the carrier is modulated, the waveshape will assume its trapezoidal pattern. Adjust the horizontal gain to give a readable presentation.

Count the number of graticule divisions at both ends of the wedge. Let **a** be the larger side and **b** the smaller side. Substitute in the formula

$$\text{Modulation percentage} = \frac{a-b}{a+b} \times 100$$

For example, assume that distance **a** is 2 graticules and distance **b** is 5 graticules.

$$\frac{5-3}{3+5} \times 100 = \frac{2}{8} \times 100$$

$$0.25 \times 100, \text{ or } 25 \text{ percent}$$

Keep in mind that this is a Lissajous pattern, plotting one voltage against another, and any phase difference will distort the edges of the trapezoid. When a phase difference exists, the sides of the wedge will be curved. Also, if the modulating signal is not a sine wave, a series of bright vertical bands will appear in the display.

In some cases, a distorted trapezoidal pattern will signal transmitter trouble. For example, in Fig. 9-17A, excessive

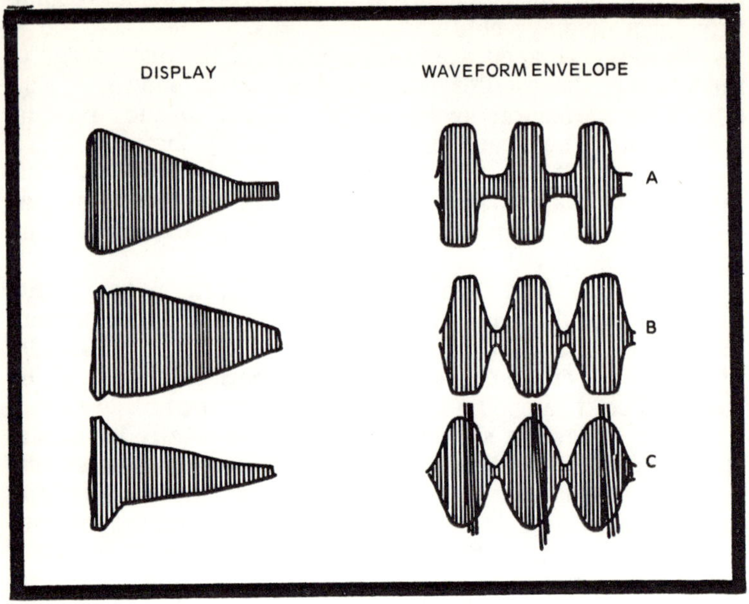

Fig. 9-17. Distorted modulation displays.

supply voltage applied to the rf power amplifier is causing overmodulation and clipping of the peaks. This results in a tail on the trapezoid and the rounding of the high end of the wedge, which represents the modulation peaks. Waveform B shows the results of too little supply voltage in the same amplifier. Notice that overmodulation does not occur, but due to the low supply voltage, the modulation peaks are clipped.

When the power amplifier is not properly neutralized, the circuit will tend to oscillate. Light and dark vertical bars will appear in the modulation envelope or trapezoidal pattern.

POWER SUPPLIES

Since power supplies produce dc voltages, why use a scope? In deriving the dc output, usually from an ac supply, the power supply forms various waveforms. These waveforms can give clues as to the power supply's condition. Also, faults in the circuit to which power is furnished can have an effect on the shape of the voltages in the power supply.

For an example of the waveforms present in a typical power supply, see Fig. 9-18. Assume for a moment that capacitor C is removed from the circuit. If X1 is a perfect rectifier—one in which there is no reverse current—the current through the resistor is a series of positive pulses occurring at a 60 Hz rate.

Now, if the load resistor, R, is removed from the circuit and replaced with a large capacitor, a pure dc voltage will appear across the capacitor. The capacitor will start to charge as the diode conducts, and after a few cycles, it will be completely charged and a steady dc will be present. If the capacitor is sufficiently large, it cannot discharge between the peaks of the pulsating dc voltage.

If the resistor is replaced in the circuit, the capacitor will begin discharging through it on alternate half-cycles, and the voltage will drop. The voltage cannot drop to zero, because a new voltage peak appears 60 times a second to recharge the capacitor. This action results in a pulsating dc output. The frequency of the pulsating dc is called the **ripple** frequency, and the ratio of the rms value of the ripple to the dc output voltage is the **ripple factor**. That is

$$K_r = \frac{E_r}{E_{dc}}$$

where Kr = ripple factor
Er = rms value of the ripple voltage
Edc = average value of the dc output voltage

Due to the capacitor storing energy between pulses, the output voltage can never fall to zero. The average output voltage with the RC filter, then, is greater than with only a resistive load. If the load is increased—that is, if the resistance of R is reduced—a greater current is drawn by the load, and the average, or dc, voltage decreases. At the same time, the ripple factor increases.

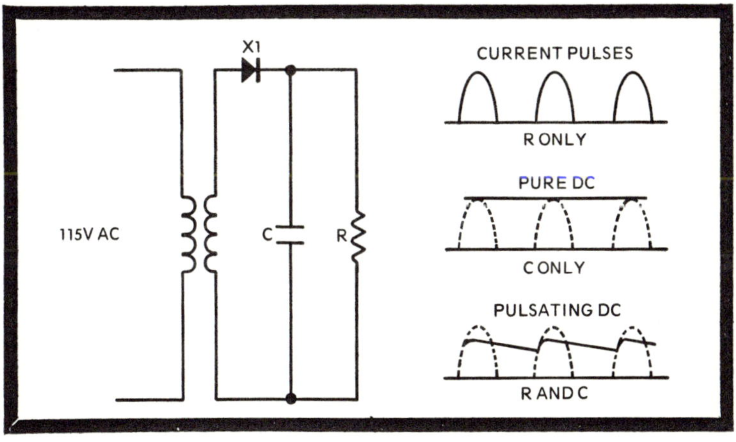

Fig. 9-18. Basic power supply waveforms.

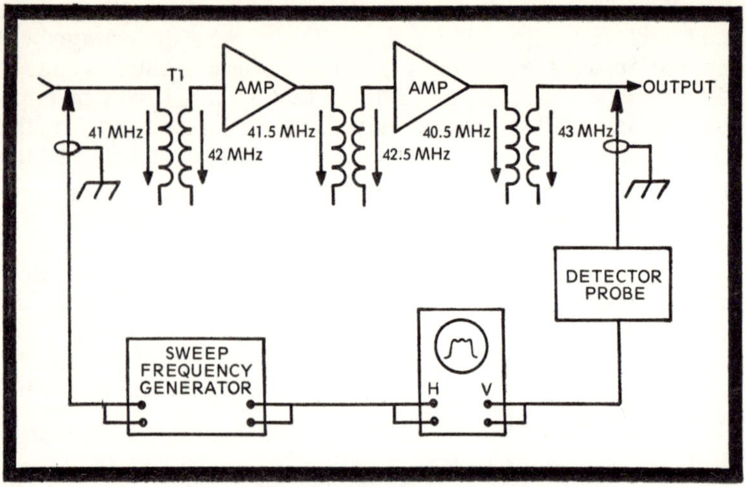

Fig. 9-19. Alignment of i-f amplifiers.

The amount of ripple resulting from a load placed across a single-capacitor filter depends upon the size of the load and the size of the capacitor. The larger the capacitor, the greater its charge. It will retain a greater charge when a load is placed across it. The greater the load current, the greater the output voltage drop and the greater the ripple.

Basically the same conditions exist in a full-wave rectifier circuit, except the frequency of the ripple is twice that in the half-wave circuit. This higher frequency aids in filtering and produces a much smoother dc; but a ripple is still present, and the amount of ripple is determined in the same manner.

Oscilloscope observations of the power supply can tell you the amplitude of the dc, of the ripple, or of the combined output. To observe the combined output, ripple and dc, set up the scope for a dc vertical input voltage and sync the scope to the ripple frequency. This will be either 60 or 120 Hz, depending on whether a half-wave or full-wave circuit is being used. A comparison of the ripple factor and dc output level with the manufacturer's specifications will give you a good indication of whether the supply is operating properly and whether it is being overloaded.

FREQUENCY RESPONSIVE CIRCUITS

One of the fastest methods of troubleshooting tuned circuits such as those found in radio and television receivers is to align the tuned circuits using a swept-frequency generator. As previously explained, the generator supplies an output that is constantly changing in frequency. This signal, varying back and forth across a center frequency, is fed to the rf and i-f amplifiers of the receiver. The output is detected and displayed as a response curve on the oscilloscope.

The diagram in Fig. 9-19 illustrates the hookup for frequency response observations in an i-f amplifier chain. The overall function of the i-f circuit is to amplify a band of frequencies centered on the intermediate frequency. The swept-frequency generator supplies the necessary signal to the input of the first i-f amplifier and a sync pulse to the oscilloscope. The oscilloscope is set up for external horizontal sync or external horizontal sweep, depending on the particular scope and generator involved, and a detector probe is attached to the vertical input terminals. As the input test signal is swept back and forth across the frequency band, a sync pulse is supplied to the scope at the start of each sweep. The output of the amplifiers is then sampled by means of the detector probe, and the resultant waveform is observed on the screen.

The displayed waveform will be a response curve showing the overall relative gain of amplifier chain, very much like the curve obtained by plotting the output voltages as shown in Fig. 5-19. Most modern signal generators used for this type of alignment also supply frequency markers in the test signal. The markers are displayed as a "pip" or "birdie" on the response curve, and enable the technician to judge the frequency response of the amplifier. As shown in Fig. 9-20, the response curve plotted by the oscilloscope is a waveform showing signal output plotted against frequency.

In this particular example, a response curve is shown centered at approximately 42 MHz. From the markers, we can judge that this amplifier has a bandwidth of about 1.5 MHz, since bandwidth is taken at the 3 dB point of the response curve (41 and 42.5 MHz here). Whether this is a "good" curve or not will depend on the frequency at which the i-f amplifier is to be operated.

Assume that the response curve in Fig. 9-20 is obtained during the tests illustrated in Fig. 9-19. The shape of this

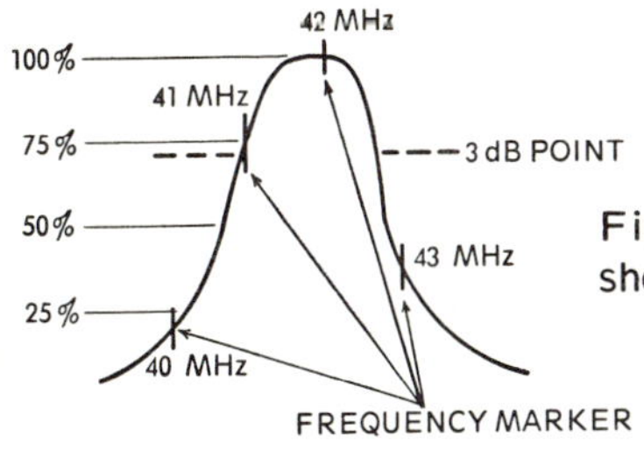

Fig. 9-20. Response curve showing frequency markers.

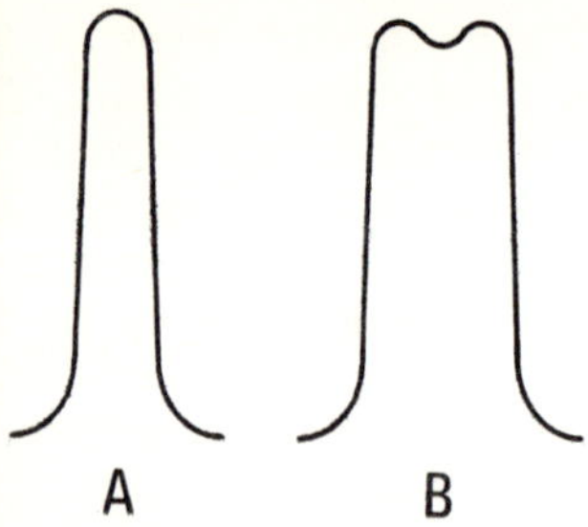

Fig. 9-21. Single- and double-peaked response curves.

curve, and thus the overall bandpass of the amplifiers, can be modified by adjusting the primaries and secondaries of transformers T1, T2, and T3. By adjusting the transformers, the degree of coupling between the primaries and secondaries and, therefore, the operating frequency, is controlled. Various waveshapes can be obtained by these adjustments.

To see the effects of a tuned circuit on response, we need to examine the operation of a single-tuned circuit and its associated response curves. Figure 9-21A illustrates the response curve for such a circuit. This type of response is typical of a very narrow-bandpass amplifier with a high gain. Although this type of response provides very good selectivity and gain, the bandpass is very narrow—not exactly what is needed in a television i-f amplifier. Waveform B shows the curve for a circuit that has been **double tuned**. The double-peaked waveform is the result of a circuit design that provides a voltage peak just above and below the resonance frequency. The resonance frequency is represented by the dip between the peaks. The purpose of this type of tuning is to increase the bandwidth—as indicated by the broader response curve—by increasing the amplitude of a number of frequencies adjacent to the center frequency. The greater the dip between the two peaks, the greater the coupling between the primary and secondary of the transformer; too large a dip is undesirable. A flat-topped curve is ideal, because all frequencies within the bandwidth of this type of amplifier will receive the same amount of amplification.

The terms **overcoupled** (close coupling) and **undercoupled** (loose coupling) relate to the degree of coupling between the primary and secondary windings of the tuned transformers. The spacing is rigid, therefore, the degree of coupling is adjusted by varying the inductance of the transformer coils with slugs.

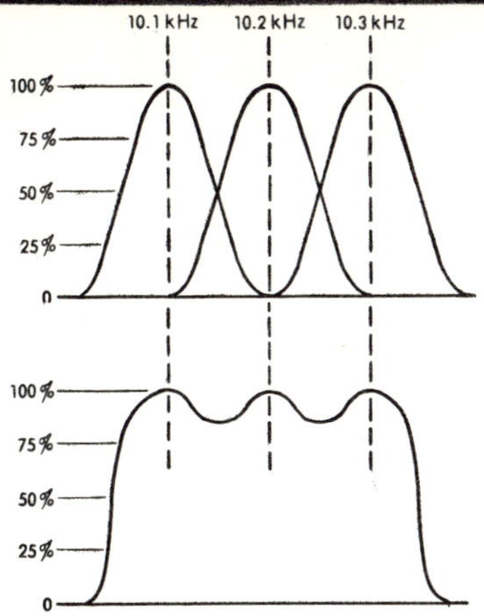

Fig. 9-22. Stagger tuning.

If a transformer is overcoupled, it will have a wide bandwidth. On the other hand, undercoupling will result in a sharply peaked waveform, representing a very narrow bandwidth and high selectivity.

To achieve the wide bandwidth and good selectivity required in radio and television receivers, two types of tuned circuits are employed, overcoupled and **stagger-tuned** circuits. The term "stagger tuning" refers to the tuning of a series of stages to slightly different frequencies. For example, in Fig. 9-22 three tuned circuits have been tuned to frequencies 1 kHz apart. The composite waveform is a triple-peaked response curve. Since the sides of the curve are quite steep, good selectivity and a bandwidth of more than 3 kHz are provided.

Assume that the circuit in Fig. 9-19 is to be response tested. With the oscilloscope and signal generator set up as shown, a multipeaked response curve can be observed as the transformer tuning adjustments are made. The primary of T1 is marked "41 MHz" and the secondary "42 MHz." The T1 slugs should shape the top portion of the leading edge and the flat top of the response waveform. Other adjustments have similar effects on the shape of the waveform. You can determine whether a particular stage is functioning properly by noting the effect of these adjustments on the overall waveshape.

FM ALIGNMENT

To align an FM receiver, a swept-frequency generator is connected to the mixer input of the receiver as shown in Fig.

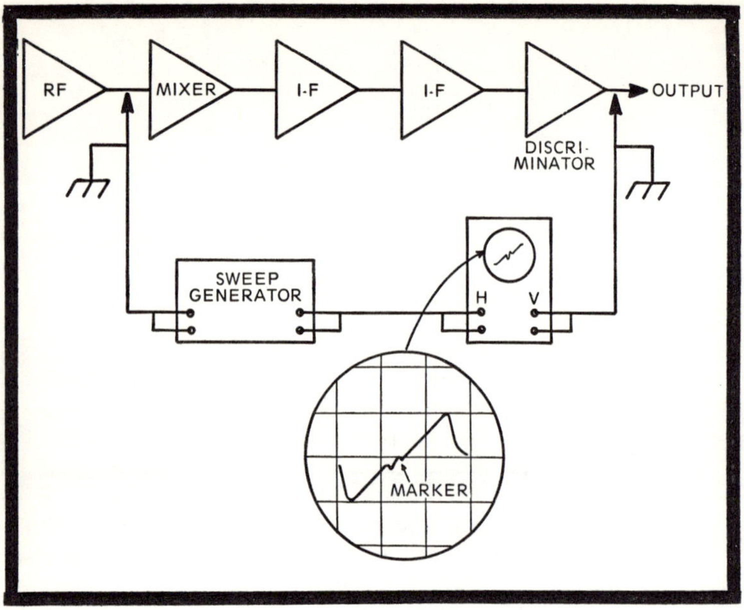

Fig. 9-23. Alignment of FM discriminator.

9-23. The center frequency of the swept output should be set at 10.7 MHz. Connect the output of the generator to the **external horizontal input** jack of the oscilloscope, and set the scope controls for external horizontal sweep. Connect the vertical input probe of the scope to the demodulator input of the FM receiver. Adjust the **vertical** and **horizontal gain** controls for a display similar to that shown. Now set the marker generator to exactly 10.7 MHz. The marker pip should appear at the center of the curve. Now adjust the i-f amplifiers according to the manufacturer's specifications to obtain the desired band-pass curve.

Once the i-f amplifiers have been aligned, move the probe to the demodulator output. The S-curve should be displayed and the 10.7 MHz pip should appear exactly in the center. Adjust the demodulator according to the manufacturer's instructions so that the marker moves equal distances from the center as the marker frequency is increased and decreased equal amounts from 10.7 MHz.

Observing Television Waveforms

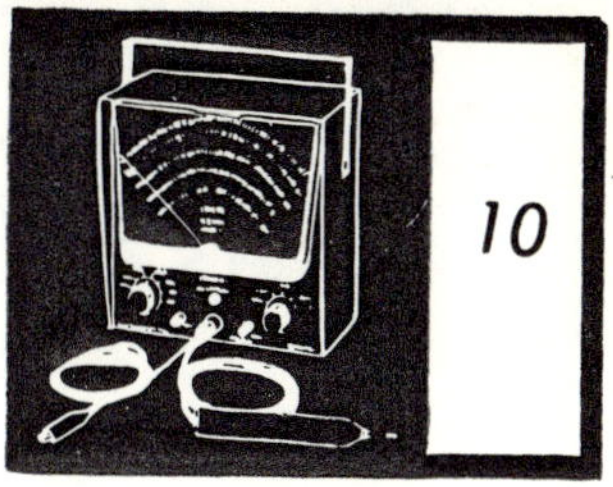

Either of the two oscilloscopes explained earlier can be used for television servicing. Of course, the triggered-sweep and sync features on the triggered-sweep model make television servicing easier. However, successful troubleshooting depends more on the knowledge of the technician and his ability to use test equipment than on his equipment. To be proficient with any scope in any troubleshooting situation, you must be familiar with the various features and performance characteristics of the scope. There's only one way to gain this familiarity—experience. Sometimes experience can even make up for lack of performance on the part of the test equipment.

VIDEO WAVEFORM ANALYSIS

Probably the most important waveform in television servicing is the composite waveform consisting of the video signal, color information, blanking pedestals, and sync pulses. Composite video signals can be observed in various circuits in the television receiver to determine whether circuits are performing normally. The two most commonly observed characteristics of the composite waveform are the horizontal scan line (looking at one complete cycle of the television horizontal scan) and the vertical frame (looking at one complete frame of the television picture).

Observing Lines and Frames

Figure 10-1 illustrates how the B&K 1460 scope with its special sync features is used to observe two complete horizontal scan lines. After tuning the television receiver to an active local channel, set the **sync** switch to **TVH** for horizontal pulse observation or to **TVV** for vertical pulse observation. Set the **triggering-source** switch to **int**, the **trig level** control to **auto**, and adjust the **intensity** control for a trace. Move the **ac-gnd-dc** switch to the ac position and connect the probe to the **vert input** jack. Connect the ground clip of the probe to the

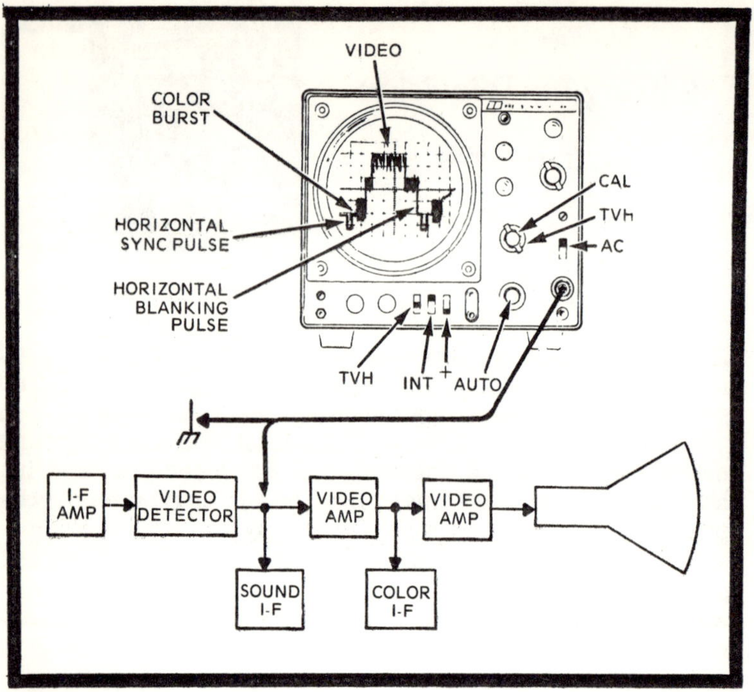

Fig. 10-1. Observing the horizontal sync pulses. (Courtesy B&K Dynascan Corp.)

television chassis. With the probe set for 10:1 attenuation, connect the probe tip to the output of the video detector.

Set the **voltage/cm** control for the largest vertical deflection without putting the display off the scale. Set the **triggering-slope** switch as described earlier. Now turn the **stability** control counterclockwise until no sweep appears on the **crt**. Then, increase the **stability** setting (clockwise) until the sweep reappears. This should result in a very stable waveform presentation. After setting the **intensity** and **focus** controls to achieve the desired brightness and sharpness, you are ready to observe the waveform.

To view a specific position of the waveform, such as the sync tip and color burst, pull the **horizontal position** control out for x5 magnification, and adjust the position of the display by rotating the **horizontal position** control.

Composite video waveforms may be checked at other points in the video circuits by moving the probe tip to those points and changing the **voltage/cm** control setting as required to keep the display within the limits of the scale, and by readjusting the **stability** control to maintain stabilization. The polarity of the observed waveform may be reversed when moving from one monitoring point to another. Therefore, it is important to remember that best synchronization of the observed waveform is obtained if the **triggering-slope** switch is used as outlined in Chapter 7.

Other wideband oscilloscopes can be used for this observation. If your scope is equipped with special television sync features, as the EICO 460 is, syncing the scope with the sweep is a simple matter of switch setting. Otherwise, set the sweep speed of your scope to about 7875 Hz to display two complete horizontal scan lines.

To observe the vertical waveforms, a similar procedure is used, as shown in Fig. 10-2. Again, if your scope is not provided with circuits that aid in syncing the scope with the television waveform, you must manually sync the scope to the waveform. By setting the scope sweep speed to 30 Hz you will be able to observe two complete vertical traces. On the B&K 1460, just change the setting of the **sync** switch to **TVV** and set the **sweep time—cm** control to the **TVV** position.

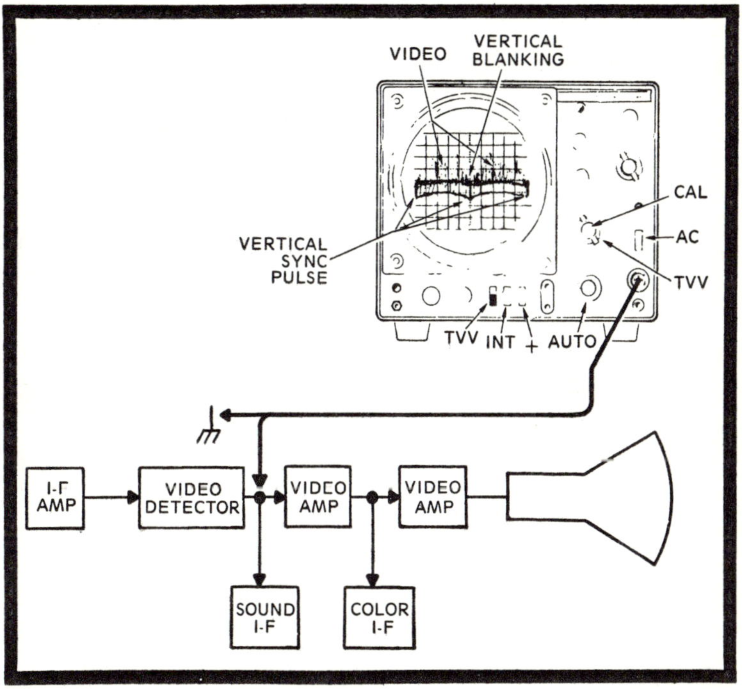

Fig. 10-2. Observing the vertical sync pulses. (Courtesy B&K Dynascan Corp.)

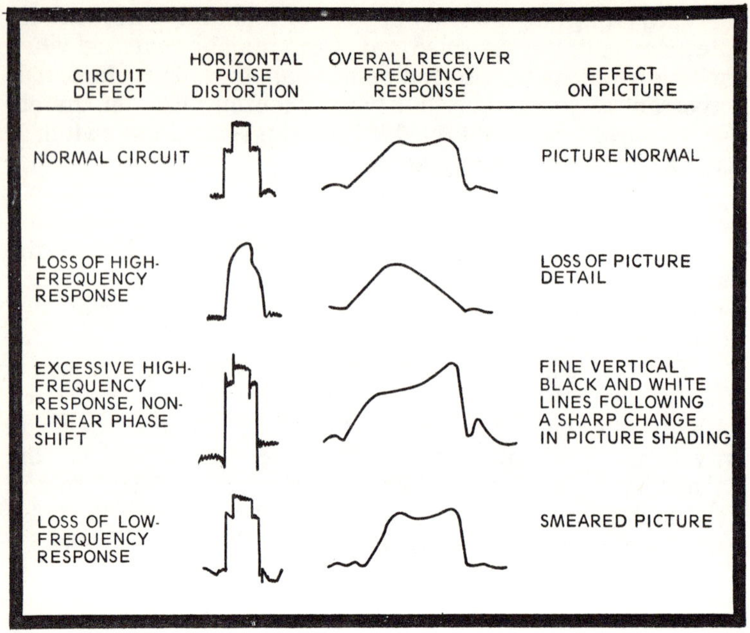

Fig. 10-3. Effects of improper circuit performance on the horizontal sync pulse. (Courtesy B&K Dynascan Corp.)

Knowledge of waveform composition, the appearance of normal waveforms, and causes of various abnormal waveforms all help you locate and correct many circuit problems. You should study the waveforms in a television receiver known to be in good operating condition, noting the waveforms at various points and comparing them to those shown in the manufacturer's service literature.

Evaluating Amplifier Response

The i-f amplifier response and general performance of the rf and i-f stages of a television receiver can be evaluated to some extent by careful observation of the horizontal sync pulse waveform as it appears at the output of the video detector. The general appearance of the waveform is affected by the performance characteristics of the rf and i-f circuits. Some examples are shown in Fig. 10-3. In Fig. 10-3A, normal operation is depicted; the picture should be good. Notice the well defined appearance of the sync pulse and the shape of the receiver bandpass curve.

The fine details of the televised scene are made up of the high-frequency components. Any loss of high-frequency response will, therefore, degrade the picture detail. In B, the bandpass curve drops off sharply toward the high end of the band, indicating a definite loss of high-frequency performance. Notice how this affects the horizontal sync pulse. Since the pulse is a rectangular waveform, we can expect to see a rounding of the leading and trailing edges. In severe cases, it will be hard to distinguish where the blanking level stops and the sync tip starts.

Next the effect of **too much** high-frequency response is illustrated. There is ringing (oscillation) in the i-f stages, causing fine, vertical, black-and-white lines to trail (appear to the right of) any sharp vertical lines in the TV picture. The bandpass curve will have a definite tilt, showing the increase in high-frequency gain. The edges of the sync pulse will overshoot, and in some cases, will be followed by a dampened oscillating wave, indicating ringing.

The low-frequency components of a rectangular or square wave are evaluated by observing the flat surfaces of the pulses. A loss of low-frequency response in a television receiver would be indicated by a smeared appearance of the picture. The bandpass response curve will droop toward the center. The horizontal sync pulse of such a receiver will have a tilt in its flat surfaces.

Some i-f faults might affect the picture synchronization before they are noticed in the actual video presentation. Bias faults in the i-f amplifiers can cause the sync tips to be clipped or compressed as shown in Fig. 10-4. Such faults are usually noticed by the erratic operation of the receiver sync circuits.

VITS

Most network television signals contain a built-in test signal, the **vertical interval test signal, or vits**, that can be a very valuable tool in troubleshooting and servicing television sets. The vits can be an aid in localizing trouble to the antenna, tuner, i-f, or video section, and shows when realignment may be required. The vits is transmitted during the vertical-blanking interval. On the television receiver, it can be seen as a bright white line above the top of the picture when the ver-

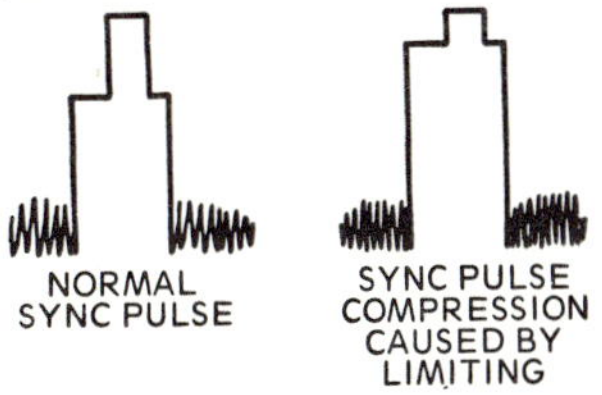

Fig. 10-4. Sync pulse compression. (Courtesy B&K Dynascan Corp.)

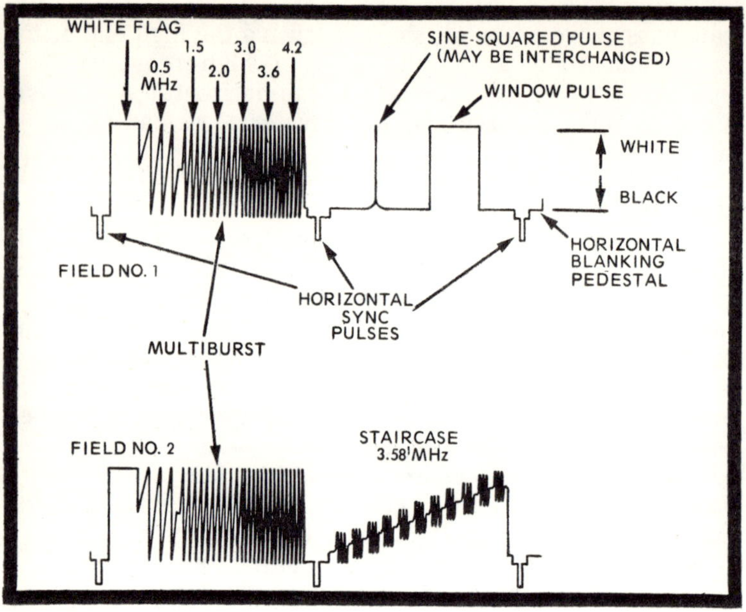

Fig. 10-5. Vits, field 1 and 2. (Courtesy B&K Dynascan Corp.)

tical linearity and **height** controls are adjusted to view the blanking bar. On sets with interval-retrace-blanking circuits, the blanking circuits must be disabled to see the vits.

VITS Composition

The transmitted vits is a precision sequence of specific waveshapes at constant frequencies and amplitudes, as shown in Fig. 10-5. The television networks use these precision signals for adjustment and checkout of the network transmission equipment, but the technician can use them to evaluate television set performance.

In the **interlace scanning** technique, the scene is transmitted in a series of vertical **frames**; each frame is composed of 525 separate horizontal lines. Two complete top-to-bottom scanning sequences, called **fields**, are used to form a frame. During the first field of any frame, the scanning beam starts in the upper left corner of the field and travels at a constant horizontal rate, causing separate horizontal traces that are a fixed distance from each other. As each horizontal trace is completed, the beam is reset to the left of the field and another scan is started. When the beam reaches the bottom of the picture, 262.5 horizontal scans have been completed. This completes the first field of one frame. Then the beam is reset to the top center of the scene, and the horizontal sweep continues. However, this time the beam is scanning between the traces of the horizontal lines in the first field.

The vits is transmitted during lines 17, 18, 279, and 280. Note that these lines would appear at the top of the picture.

The first vits transmission, which occurs in line 17, begins with a "flag" of white video, followed by sine-wave signals of 0.5, 1.5, 2, 3, 3.6 and 4.2 MHz. This sequence of frequencies is called the **multiburst**. The multiburst is also transmitted during line 279, as shown in Fig. 10-6. These are the two transmissions that are the most valuable to the technician. The other two vits transmissions, lines 18 and 280—which contain the sine-squared pulse, window pulse, and staircase of 3.58 MHz bursts at progressively lighter shading—are valuable to the network, but have less value to the technician.

Due to interlaced scanning, field 1 is interlaced with field 2, so that line 17 is followed by line 279, and line 18 is followed by 280. The entire vits appears at the bottom of the vertical-blanking pulse and just before the first line of video. Since each of the multiburst frequencies is transmitted at equal strength, they can be used as a troubleshooting aid. By observing the comparative strengths of these frequencies after the signal is processed through the television receiver, the frequency response of the set can be estimated.

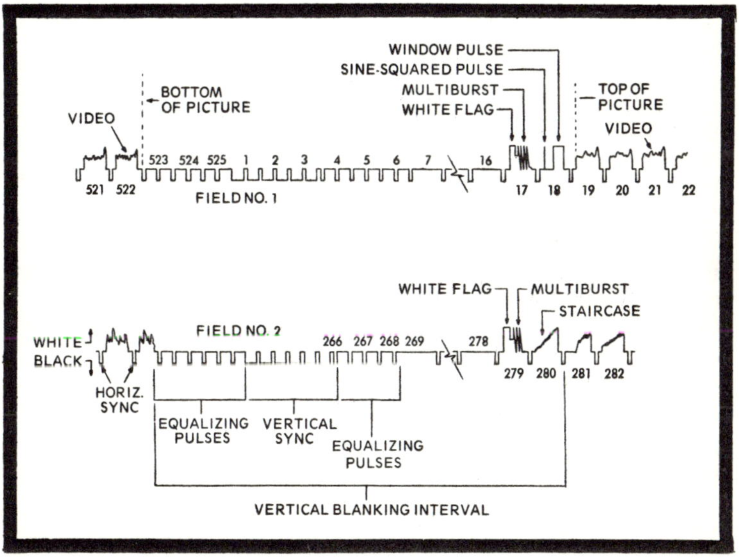

Fig. 10-6. The vertical-blanking interval, showing the vits. (Courtesy B&K Dynascan Corp.)

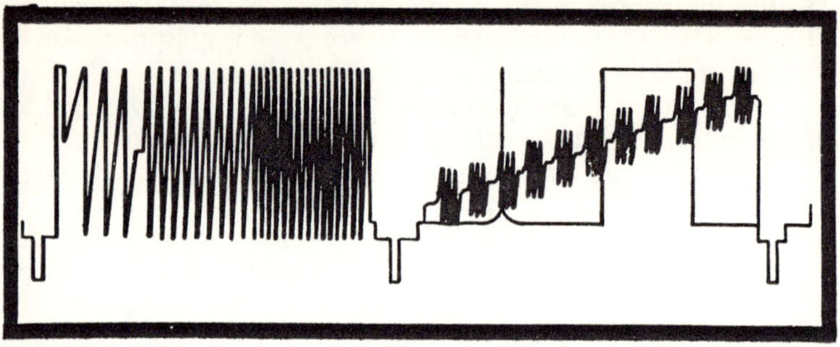

Fig. 10-7. Vits fields superimposed. (Courtesy B&K Dynascan Corp.)

Observing Vits

To observe the vits using the B&K 1460, set the probe for 10:1 attenuation and connect it to the output of the video detector. If the television receiver has a vertical-retrace-blanking circuit, follow the manufacturer's instructions for bypassing that circuit during these measurements. Set up the scope for **TVV sync** and **TVV sweep time**, as previously described, to show two vertical fields. Place the **variable-hor gain** control in the **cal** position. Now, reduce the sweep time to 0.1 msec / cm by placing the **sweep time/ cm** switch to the **0.1 msec / cm** position. This expands the display by increasing the sweep speed. The vits should appear to the right on the expanded waveform display.

Further expand the sweep by pulling outward on the **horizontal position** control so that the x5 magnification feature is employed. Rotate the **horizontal position** control in a counterclockwise direction, moving the trace to the left until the expanded vits appears. The brightness level of the display will be reduced.

The waveform should be similar to that shown in Fig. 10-7. For the oscilloscope display, each vertical sync pulse starts a new sweep. This causes line 17 and line 279, the two multiburst transmissions, to be superimposed, as are lines 18 and 280. The multiburst signals are identical, reinforcing the trace. However, lines 18 and 280 are not identical, and the signals are superimposed over each other. This should present no problem, since the multiburst is normally the signal of interest.

All frequencies of the multiburst are transmitted at the same level, but they should not be equally coupled through the receiver, due to its response curve. Figure 10-8 shows the desired response for a good color television receiver, identifying each frequency of the multiburst and showing the allowable attenuation for each. Remember that —6 dB equals the half-reference voltage (the 2.0 MHz modulation should be used for reference).

To localize troubles, start by observing the vits at the video detector. This will localize the trouble to a point either before or after the detector. If the multiburst is normal at the detector, check the vits on other channels. If some channels look okay but others do not, you probably have tuner or antenna problems. If the vits is abnormal at the video detector on all channels, the trouble is probably in the i-f amplifiers. Should the vits response at the detector output prove normal for all channels, the trouble is in the video amplifiers.

TELEVISION ALIGNMENT

Alignment of the tuners, i-f amplifiers, and chroma circuits in television receivers requires a high-quality

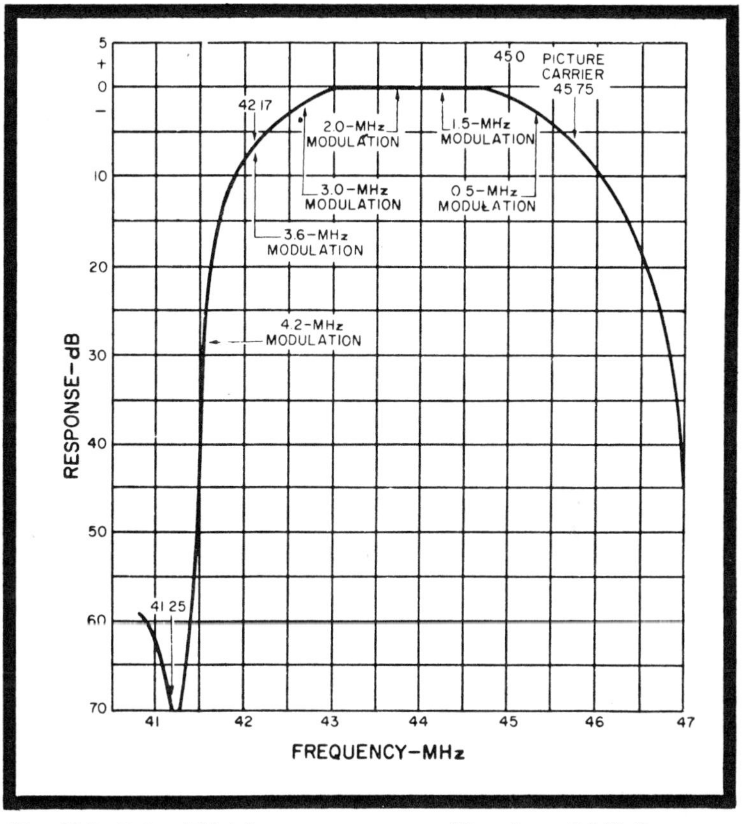

Fig. 10-8. Color TV i-f response curve. (Courtesy B&K Dynascan Corp.)

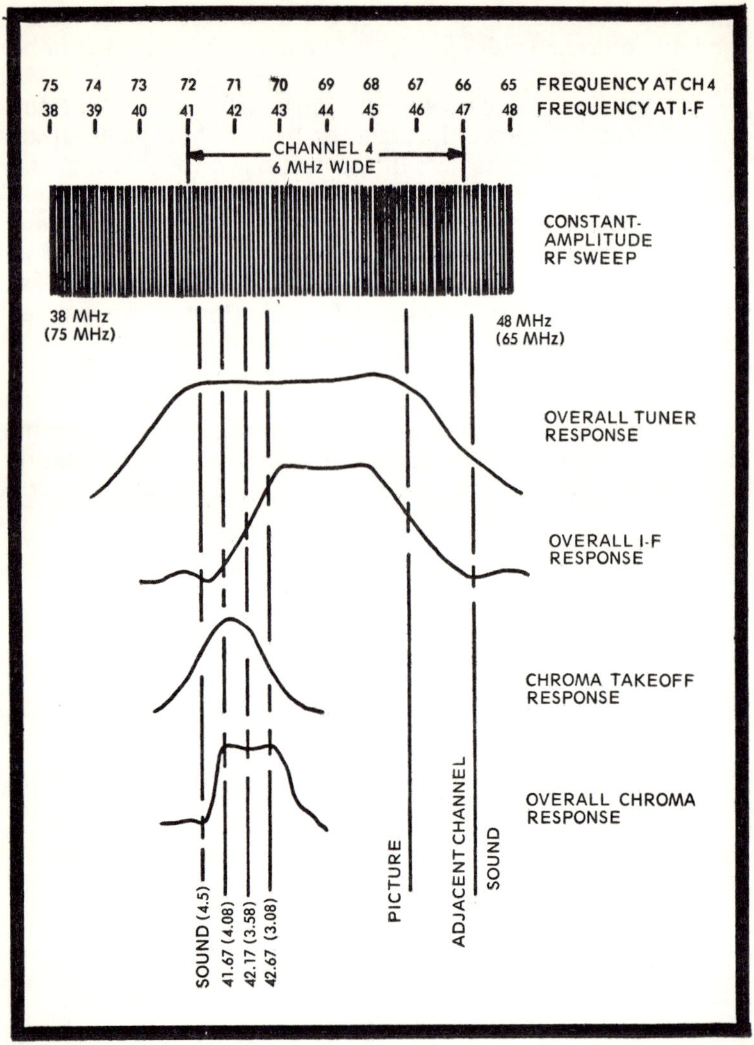

Fig. 10-9. Television response curves. (Courtesy B&K Dynascan Corp.)

oscilloscope. In addition, you will need a sweep generator for video, i-f, and rf sweep; marker generators; dc bias supplies; and an electronic voltmeter. In this book, emphasis will be placed on the proper use of the scope. For proper use of the sweep generator and other equipment, refer to the manuals for those instruments. For complete alignment instructions of any television receiver, refer to the manufacturer's service data.

Response Curves

The most rapid way to determine the overall condition of the tuner, i-f, and chroma portions of the television receiver is to provide a constant-amplitude signal which sweeps through the entire bandwidth of a given television channel repetitively at a controlled rate. As this signal is processed through the tuned portions of the receiver, it is shaped by the gain and bandpass properties of the various sections. Because the signal is channeled from one series of tuned circuits to another, it is important that each section have the proper characteristics. If the signal is demodulated at certain points and the envelope observed, the gain and bandwidth properties up to that point can be determined.

Figure 10-9 shows the sweep signal, with the basic response curves of the tuner, i-f amplifiers, and chroma bandpass circuits below it. The bandwidths are shown approximately to scale. These outlines are similar to the curves that would be obtained if the outputs of the various sections of the television receiver were demodulated and the curve observed with a scope. Some reference frequencies are identified to show the importance of proper alignment. Notice that the chroma frequencies are on the slope of the i-f response curve. This area is the most critical, because improper i-f alignment in this area will affect the amplitude and shape of the chroma response curve and this, in turn, will affect color picture quality.

Notice that the chroma information is located on a constant-amplitude portion of the transmitted television spectrum. Note also that the relative amplitudes of the chroma information are modified by passing through the tuned circuits of the tuner and i-f amplifiers. This is shown by reference to the overall i-f response curve. The signal information at the upper end of the chroma frequency range (4.08 MHz) is reduced in amplitude with respect to the signal level at the lower end (3.08 MHz). To compensate for this, a chroma takeoff coil is used between the i-f output and the bandpass amplifier of the chroma section. The chroma takeoff coil is tuned to the upper end of the chroma frequency range, usually 4.08 MHz, and provides a response as shown in Fig. 10-9. The result of combining the response of the i-f section with the response of the chroma takeoff coil is to produce a flat overall response in the chroma frequency range of 3.08-4.08 MHz.

Obtaining Response Curves

The best method of checking alignment and determining which stages require alignment is to inject a swept rf sweep signal at the antenna terminals. The agc bias line must be clamped by application of bias or by grounding. The outputs of the i-f and chroma circuits are then observed with an oscilloscope and compared to the manufacturer's recommended response curves.

By this comparison, you can determine which portions of the receiver need alignment. For example, if the i-f response is satisfactory but the chroma response is not, then the problem is between the video detector output and the bandpass amplifier output. If the i-f response and the chroma response are poor, it is likely that the i-f requires touchup, particularly if the response is poor on the slope affecting chroma response.

The rf portion of the tuner seldom creates an alignment problem, because its bandpass is so much greater than that of the i-f section. However, the mixer output circuit may require attention. This is part of the tuned matching network between the tuner and first i-f stage. A separate prealignment procedure is given for the link circuits (connecting tuner and i-f) by some manufacturers.

Once the troublesome portion of the receiver is determined, an alignment check of that section can be performed. Some manufacturers suggest using signal combinations at the tuner antenna terminals which can generate i-f and video sweep frequencies in the receiver, so that overall alignment can be done by selecting the right combination of input signals. One way of doing this is to first connect an rf sweep generator for i-f alignment. After this is complete, the picture carrier frequency for the channel being used is selected and this is modulated by a video sweep signal. The video sweep modulation is recovered at the video detector of the television receiver and applied to the chroma bandpass circuits for alignment of these stages. This is the **video sweep modulation,** or vsm, method.

Some manufacturers recommend the use of an i-f sweep frequency injected at the mixer for i-f alignment. The i-f picture carrier frequency at 45.75 MHz is then modulated with a video sweep voltage (vsm again). As before, this is detected at the video detector of the television receiver, and the recovered sweep voltage is used for the chroma circuit alignment.

Another method is to first video-sweep align the chroma circuits directly. The i-f is then aligned, and the vsm of the i-f picture frequency (45.75 MHz) is used to check the combined effect of i-f and chroma alignment on the chroma response. Usually, a touchup of the chroma circuits is done to obtain the desired final overall chroma response.

Using the Scope in Alignment

In conjunction with i-f alignment, practically all manufacturers recommend pretuning i-f traps by injecting specific frequencies, usually at a specified tuner test point. In all cases, the manufacturer's method is the best for his particular receiver. The following general procedures are given to acquaint you with scope operation during alignment.

The tuner alignment of a television receiver is accomplished with the B&K 1460 by connecting the output of the sweep generator to the antenna terminals of the television set (as shown in Fig. 10-10) and adjusting the generator to sweep one channel. Tune the set to that channel and set the vertical controls of the scope for maximum sensitivity, so that the generator can be operated at a low level to avoid distortion brought about by overloading.

Connect the ground clip of the oscilloscope probe directly to the tuner shield to minimize hum pickup. Set the vertical probe to **direct** and connect it to the tuner test point. The oscilloscope sweep and sweep generator must be in exact synchronization and phase for proper presentation of the response curve. This is easily accomplished for sinusoidal or sawtooth sweeps by setting the oscilloscope for an external horizontal input and connecting the horizontal sweep voltage

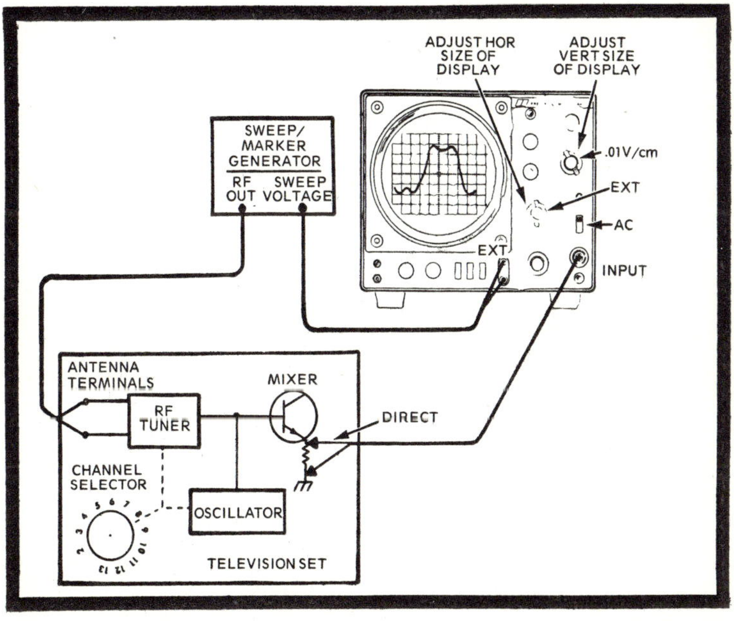

Fig. 10-10. Typical tuner alignment setup. (Courtesy B&K Dynascan Corp.)

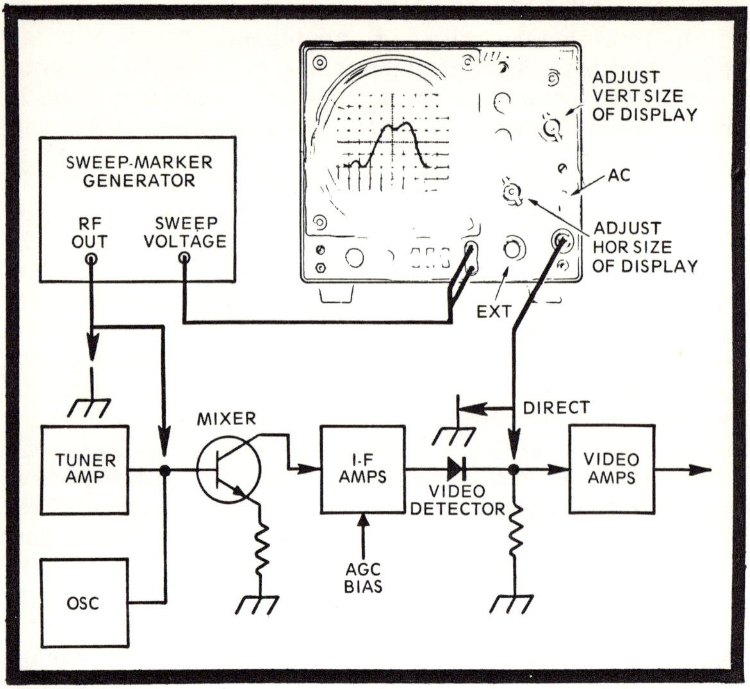

Fig. 10-11. Typical i-f alignment setup. (Courtesy B&K Dynascan Corp.)

from the sweep generator to the **external horizontal input** terminals of the scope. Select the marker generator frequencies required to measure the upper and lower response of the tuner.

The tuner response curve is now displayed on the oscilloscope. The manufacturer's instructions give the response curve specifications and the necessary adjustments for realignment.

The alignment of the i-f amplifiers, shown in Fig. 10-11, is somewhat similar. Connect the output of the sweep generator to the signal injection point of the mixer. Adjust the sweep generator to sweep the i-f bandwidth. If the tuner has been previously aligned, an rf sweep may be applied at the antenna terminals. Synchronize the oscilloscope sweep with the sweep generator, as in tuner alignment. Connect the ground clip of the oscilloscope vertical probe to the television receiver chassis and the probe tip to the video detector output.

Set the vertical gain controls, **volts / cm** and **variable**, for suitable viewing of the response curve, keeping the sweep generator output level low to prevent overloading. Be sure to follow the manufacturer's recommendations on disabling the agc. Select the marker frequencies required to check the frequencies of interest and follow the manufacturer's instructions for evaluating the response curve and making any necessary adjustments. A typical response curve is shown in Fig. 10-12.

The i-f alignment must be satisfactorily completed before starting the chroma alignment procedure. If direct injection of the video sweep signal is used rather than i-f sweep injection, the response curve is altered drastically. Make sure you follow the manufacturer's procedure exactly for such direct injection of video sweep for chroma alignment.

Leave the sweep-marker generator and agc bias connected as for i-f alignment. Set the sweep generator to sweep approximately the 41-44 MHz band of frequencies. Use the same i-f injection level that was used for i-f alignment. Apply the proper dc bias to the color killer to "enable" the color amplifiers.

Synchronize the oscilloscope sweep as before. Use a demodulator probe for the vertical input to the oscilloscope, as shown in Fig. 10-13, and measure the response curve at the input to the color demodulator circuits.

Set the vertical gain controls of the oscilloscope for a convenient viewing size. The display should be similar to that in Fig. 10-14.

SIGNAL TRACING

Signal tracing with an oscilloscope is one of the fastest methods to isolate television receiver problems. The importance of the shape of the transmitted composite signal at the output of the video detector was discussed earlier. In some instances, this broadcast signal is not as useful as one that can

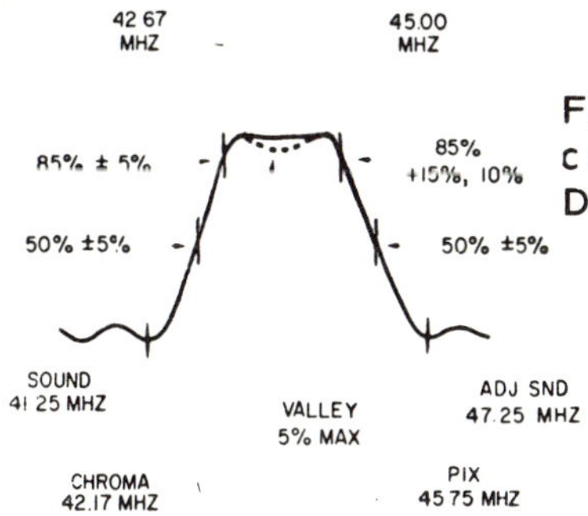

Fig. 10-12. Typical i-f response curve. (Courtesy B&K Dynascan Corp.)

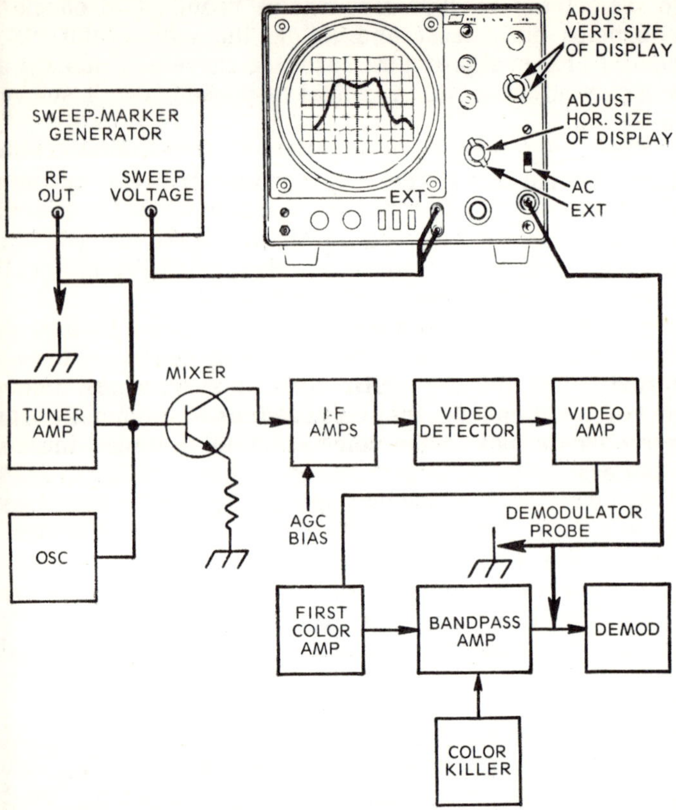

Fig. 10-13. Typical chroma alignment setup. Dynascan Corp.)

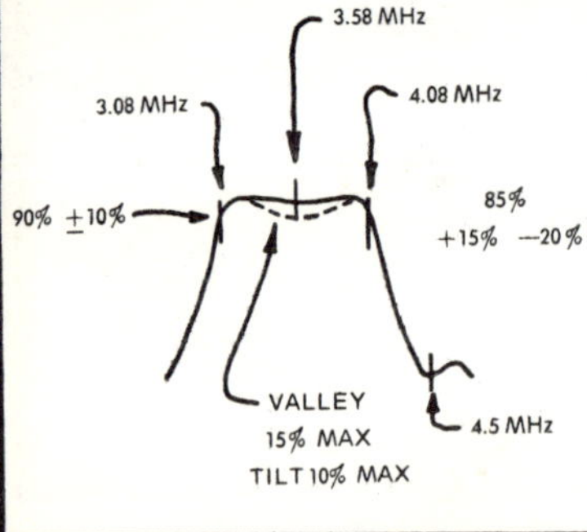

Fig. 10-14. Typical chroma response curve. (Courtesy B&K Dynascan Corp.)

be readily generated and controlled at the test bench. Since the broadcast scene is constantly changing in content, the color generator offers a more constant signal for oscilloscope evaluation. This is especially true when making waveform observations in the video or color sections of the receiver. The color generator can also be used as a signal source for video troubleshooting in a monochrome receiver. As a general rule, I use the station signal when troubleshooting the deflection section or other pulse circuits, and a color generator in the other circuits. Depending on the frequency of the generator's output, it may be connected to either the antenna terminals or the input to the i-f section.

Video Section

The signal used for waveform observation in the video section may be either a station signal or a generator signal, and may be observed at either the horizontal or vertical rate.

A block diagram of a typical video section is shown in Fig. 10-15. The output of the video detector is routed through two stages of amplification and a delay line before final amplification by the driver. The first and second video stages are

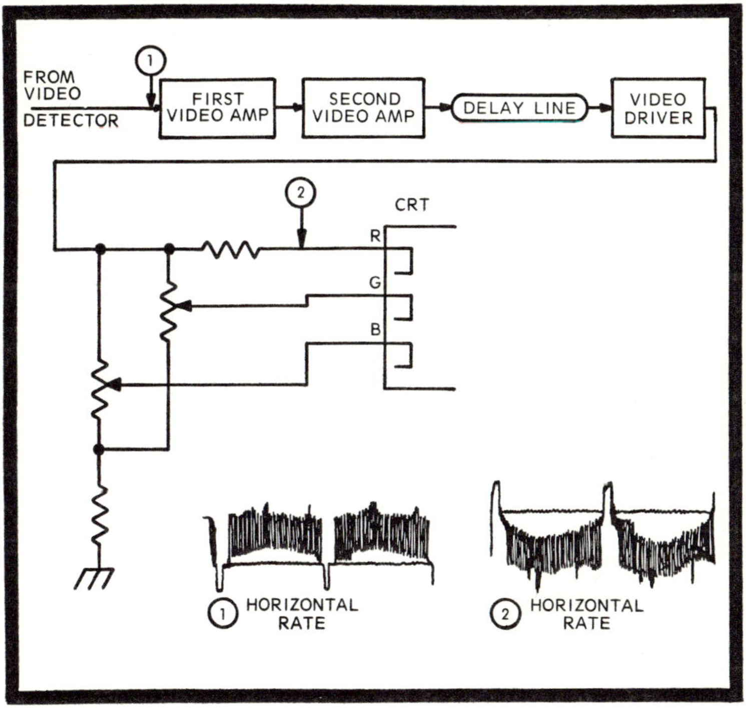

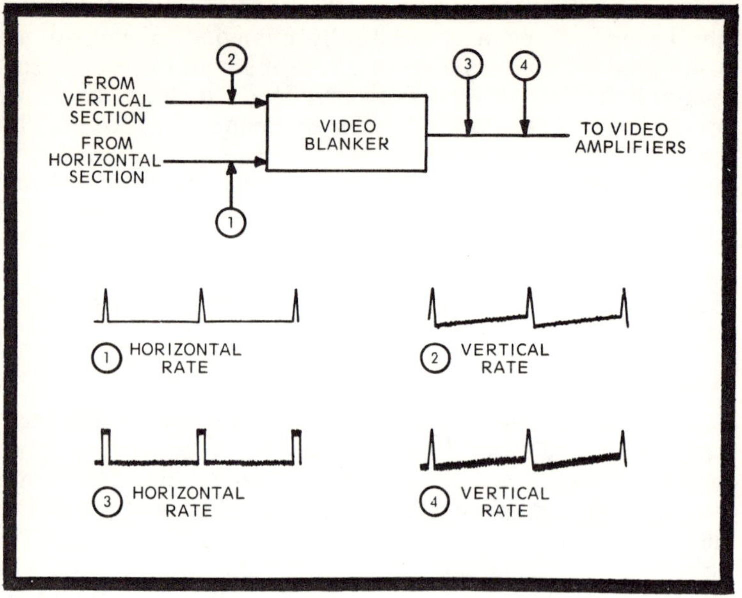

Fig. 10-16. Video-blanker waveforms.

the source of sync, agc and, chroma signals in most sets. The manufacturer's service data should be consulted for expected waveshapes.

To use the station signal as a signal source, the antenna and tuner connections must be in place. Tune the set for the best possible picture, and observe the waveshapes at the desired points using a low-capacitance probe. For displaying two complete horizontal scanning lines (as shown in Fig. 10-15), set the scope up for internal sync using the observed signal as the sync source, with the scope sweep set for 7875 Hz.

Another stage frequently associated with the video section is the **blanker**, shown in Fig. 10-16. This stage receives pulse signals during the horizontal and vertical retrace times to control the output of the video section. Usually, the output of the blanker is fed to the first or second video stage to bias that stage off during the retrace periods. Vertical signals should be observed at a 60 Hz rate, horizontal signals at a 7875 Hz rate.

Sync

Sync signals can also be observed using the station signal. Basically, the setup is the same as that for video waveform display.

The block diagram in Fig. 10-17 shows two sync inputs. A noise signal from the video detector is applied to the noise gate circuit. This signal will cause the noise gate to bias the sync separator circuit off when large noise pulses reach the video section. The sync input is routed directly to the separator circuit, where vertical and horizontal sync signals are removed for their respective circuits. The outputs of the separator should be observed at either the 60 or 7875 Hz sweep rate.

AGC

The agc section usually employs two stages, a keyer and an agc amplifier. A dc output is sent from the amplifier to the tuner and i-f amplifiers for gain control.

A keying signal in the form of a pulse is applied to the keyer from the horizontal section, as shown in Fig. 10-18. This allows the agc voltage to be referenced to the average signal strength of the received horizontal sync pulse rather than the average video signal. A video signal from the detector passes through the keyer when the keying pulse is present. Both signals should be observed at the horizontal rate.

Vertical

The vertical sync signal from the sync separator is applied to the vertical oscillator. Depending on the set, the

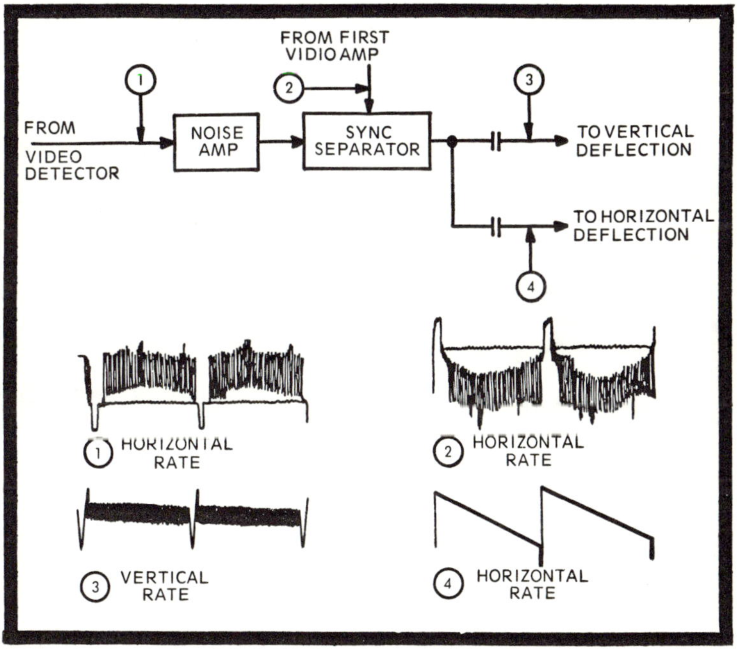

Fig. 10-17. Sync waveforms.

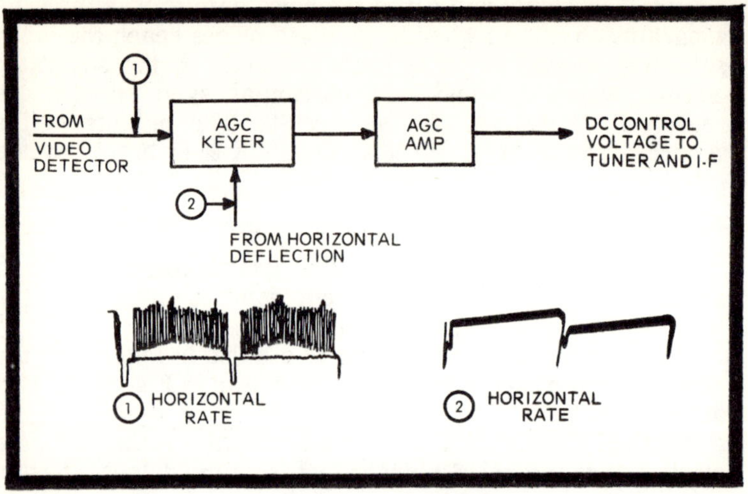

Fig. 10-18. Agc waveforms.

vertical section may consist of a signal oscillator stage, an oscillator followed by a driver (Fig. 10-19), or an oscillator followed by two or more stages of amplification. The waveforms are generally the same, and are observed at a 60 Hz rate.

Exercise care when making oscilloscope observations in the vertical output circuit. Insure that the signal being measured doesn't exceed the limits of your scope.

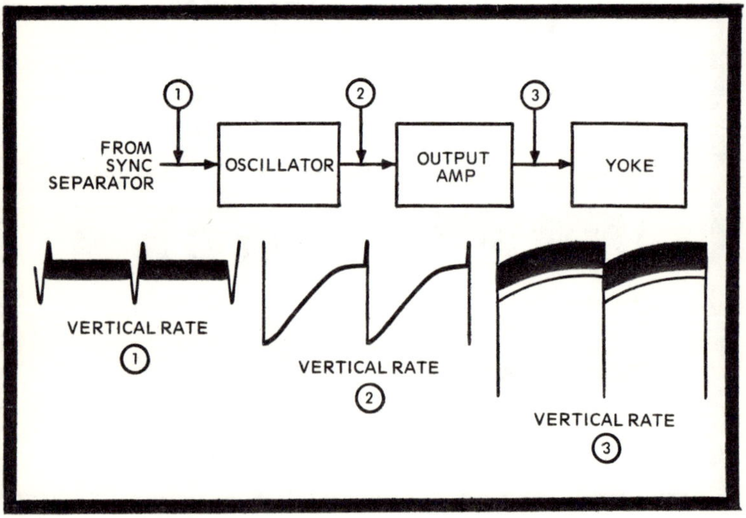

Fig. 10-19. Vertical deflection waveforms.

Horizontal

The typical horizontal deflection section consists of three stages plus the output transformer and yoke, as shown in Fig. 10-20. Again, a station signal can be used as the signal source, provided all preceding stages are in good working order.

The afc stage receives the horizontal sync signal from the sync separator and compares it to the actual horizontal output, as sampled from the output transformer. The dc output of the afc unit, representing the difference between the two signals, is used to control the frequency of the horizontal oscillator. Sweep voltages from the oscillator are amplified and shaped in the output circuit for application to the output transformer and yoke.

The oscilloscope sweep rate should be set for 7875 Hz. Again, watch out for high voltages, especially near the output circuit.

Color Signal Observations

One piece of equipment often used in color television servicing is the **keyed-rainbow** (color bar) **generator**.

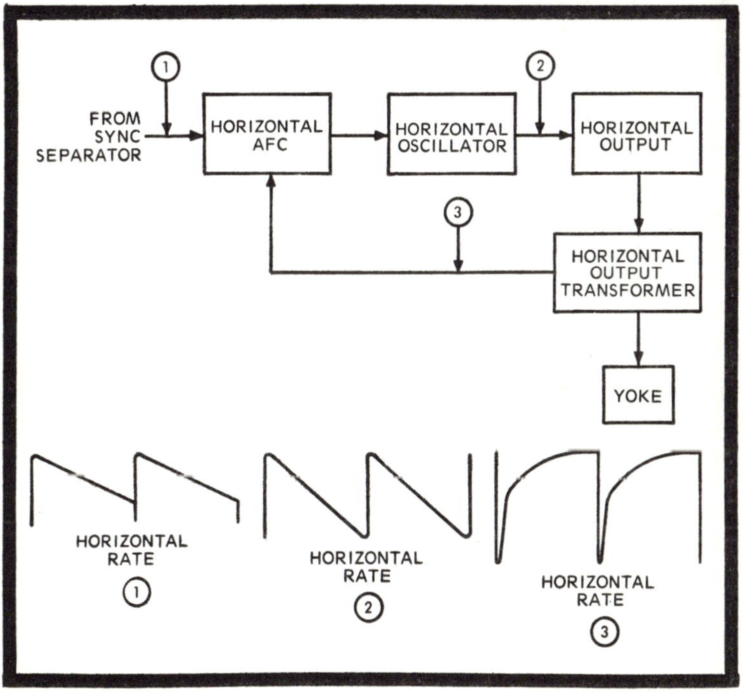

Fig. 10-20. Horizontal deflection waveforms.

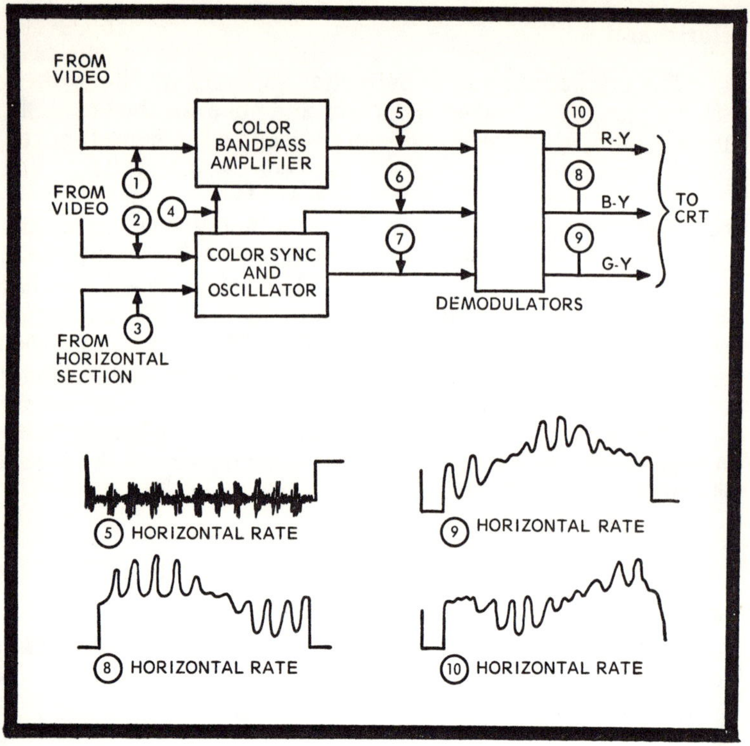

Fig. 10-21. Color waveforms.

Originally, these generators were intended to be used in such service operations as convergence and color adjustments. Added features have moved these once-in-a-while tools into the most-often-used category. For example, some of the late-model color generators have outputs at rf, i-f, and video frequencies. Couple this type of instrument with a good oscilloscope and—in my opinion—you have an unbeatable combination for general television troubleshooting. In this discussion, let's assume that the color generator has the capability to generate a 10-bar color display (similar to that in Fig. 8-4) at all the high video frequencies.

With such a generator connected to the receiver (either at the antenna terminals, i-f input, or video section input, depending on the generator), and the generator and set tuned for the best color bar display, we can determine the operating condition of the color section, using an oscilloscope to observe various waveforms. The key points are shown in Fig. 10-21, all waveforms being observed at the horizontal rate.

A portion of the video signal is fed to the color sync circuits along with a pulse from the horizontal section. From these inputs, three outputs are derived. One, a gating signal for the color bandpass circuit, is used to eliminate the bandpass output during monochrome transmissions. The other two outputs are the 3.58 MHz local oscillator and quadrature signals.

The bandpass amplifier receives, in addition to the killer bias from the sync circuits, the composite video signal from the video section. Bandwidth limitations of the amplifier limit the output to the color signal contained in the composite signal. The color components are routed to the demodulators along with the 3.58 MHz and quadrature signals. The output of the demodulator is the R-Y, B-Y, and G-Y signals for the crt.

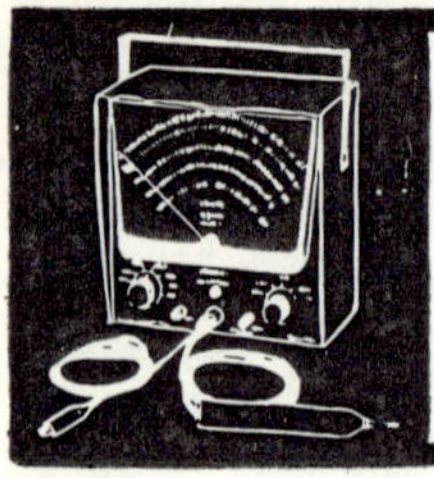

11 Checking Scope Operation

When I first started to use a scope, I had considerable doubts about the waveforms I observed. Was the distortion I encountered caused by a fault in the circuit being tested, or was it the result of a faulty scope? Many times I found that it was neither. Poor hookups caused much of the trouble. Always check your setup, especially when working in high-frequency circuits. The slightest bit of stray capacitance in leads can cause serious waveform distortion. Always use the shortest test leads possible, to prevent adding any unnecessary loading to the circuit under test. After the setup has been made and you have a waveform displayed, touch the case of the scope and run your hand along the test leads. Any change in the waveform display on the scope means added grounding is needed. Make all ground connections firmly, and occasionally check the shielding and ground clips of your test leads.

If you have an actual service problem with your scope, there are several ways you can handle it. The first reaction is usually a do-it-yourself response. This is good in most cases, as long as it is not taken too far. If you are a hobbyist and your scope is a fairly simple one, you can probably afford to take the time that it requires to troubleshoot and repair the unit. But if you run a large service department equipped with expensive calibrated equipment, you may want some outside help. It all depends on your particular needs and the particular equipment. In many cities, there are independent companies that specialize in repair and calibration of test equipment. Of course, there are the local representatives of the manufacturer, and the factory repair service offered by most test equipment manufacturers. If your scope is a recent one, it may still be covered by a warranty.

In any case, make sure that you actually have a problem with the scope and that it's not just a relatively simple adjustment that's causing the problem. Here are some tests that you can make of your scope.

TESTING THE VERTICAL PROBES

There are several types of vertical probes used with the oscilloscope: demodulator, low-capacitance, direct, and combination probes. The direct probe has very few features to be tested. An occasional check of the shielding and ground clips is about all that's necessary. The best check of a probe is substitution. When you find a faulty probe, troubleshoot it just as you would a similar circuit in a radio or television receiver.

Demodulator Probe

The demodulator probe shown in Fig. 11-1, under typical operating conditions, receives a modulated rf input and furnishes a low-frequency detected signal (carrier envelope) and a dc proportional to the amplitude of the rf carrier. This circuit is similar to the detector of an AM receiver.

To test this probe, use the setup shown in Fig. 11-1. Set the scope up for routine waveform display, syncing internally on the probe signal. Set an rf generator up for an output on the

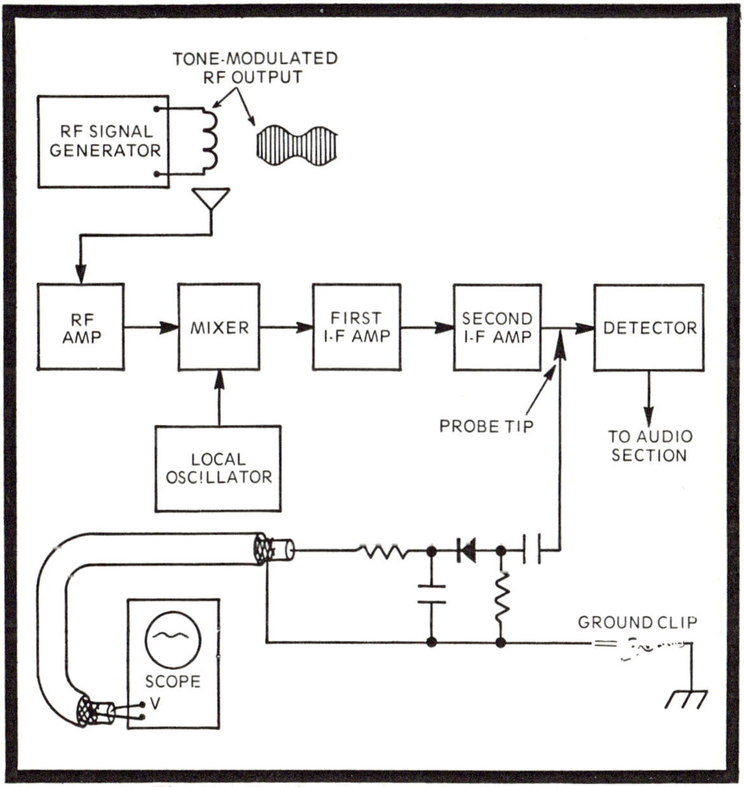

Fig. 11-1. Testing the demodulator probe.

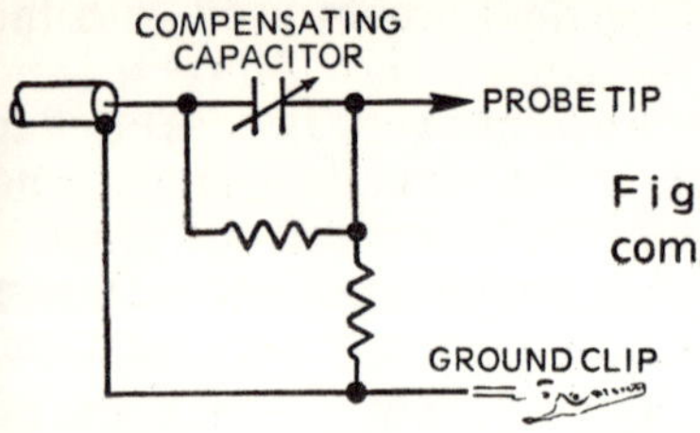

Fig. 11-2. The frequency-compensated probe.

AM broadcast band, and modulate the output with the tone modulation signal in the generator (usually 400 Hz). Using a loop probe on the generator, loosely couple the modulated rf to the antenna of an AM broadcast receiver. Tune the receiver to the generator signal, obtaining a steady 400 Hz tone at the speaker. Connect the demodulator probe to the vertical input of the scope and attach the probe ground clip to the chassis ground of the receiver. Now, touch the probe tip to the output of the final i-f amplifier of the receiver, and sync the scope on the signal to display several complete cycles of the modulating signal.

If the waveform is not present or is distorted, troubleshoot this circuit just as you would any demodulator. An ac voltmeter can be used to follow the signal through the probe, and an ohmmeter can be used to check the diode and resistors. Don't forget the probe cable itself; check it for continuity—both the shield and signal-carrying conductor. Also check for shorts between the signal-carrying conductor and the shield.

Low-Capacitance Probe

The low-capacitance probe (Fig. 11-2) attenuates the signal being observed by the same factor by which it multiplies the input impedance of the scope. For example, if the impedance of the probe is 33M and the input impedance of the scope is 3.3M, the probe will attenuate the signal by a factor of 10. If you do not know the attenuation factor of this probe, it can be measured once you know that the probe is operating correctly. Usually, the only maintenance required of this probe is to check the condition of the clips and shielding and the adjustment of the probe's frequency response. One manufacturer recommends the following procedure to compensate the probe.

Set up a square-wave generator for an output between 10 and 50 kHz. Connect the probe to the **vertical input** terminal of the scope and attach the probe to the generator output. Sync the scope to the generator signal and display two or three complete cycles of the input. Observe the presentation on the crt. The purpose of the probe is to furnish an output that is identical in shape to the input.

If the probe is correctly compensated, the square wave should appear as waveform A in Fig. 11-3. Undercompensation will cause the leading edges of the waveform to be rounded off (waveform B) while overcompensation results in overpeaking (waveform C). Using an insulated alignment tool, adjust the tuning capacitor of the probe for the correct waveshape.

Some scopes have a sawtooth output available at either front- or rear-panel terminals. If a square-wave generator is not available, this sawtooth voltage may be used to compensate the probe. To use this test voltage, which is taken from the scope's horizontal sweep circuit, set the scope sweep to about 1000 Hz. With the probe attached to the **vertical input** terminals, touch the probe tip to the sawtooth waveform terminals. Adjust the **vertical** and **horizontal gain** controls for a display that covers approximately two-thirds of the crt. A correctly compensated probe will give a waveform similar to that of waveform A in Fig. 11-4. Undercompensation and overcompensation will cause the waveform to be curved, as shown in B and C. The probe should be adjusted for the correct waveshape, in the same manner as when the square-wave test signal is used.

If the probe cannot be adjusted for the correct waveform, it may be due to poor frequency response in the scope or a poor waveform output from the generator, rather than a fault in the probe itself. The generator's output can be observed directly with a scope known to be in good operating condition and, later in this chapter, you'll see how the frequency response of the scope itself may be checked.

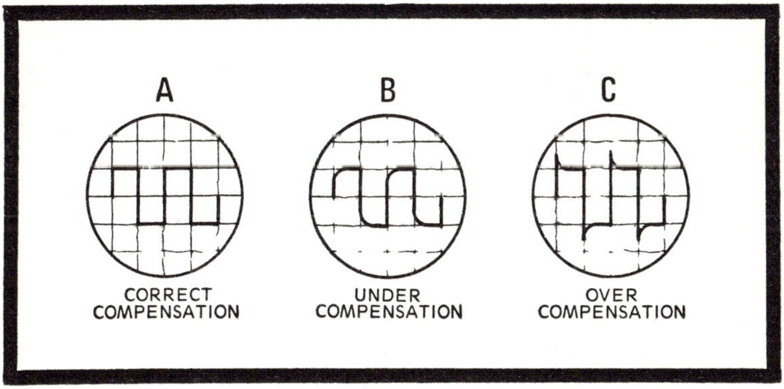

Fig. 11-3. Using a square wave to compensate a probe.

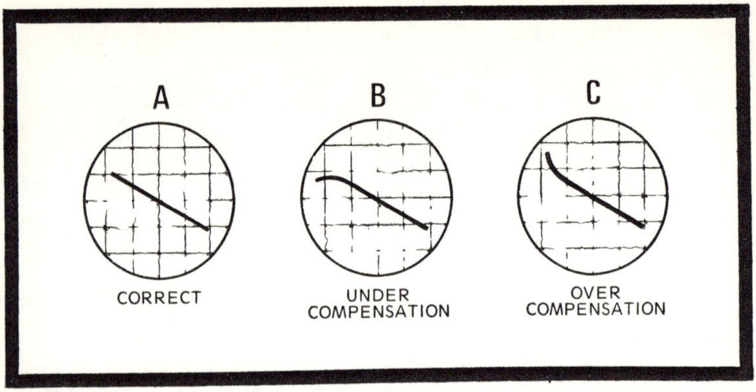

Fig. 11-4. Using a sawtooth to compensate a probe.

You can determine the attenuation in a probe if you know the probe is correctly compensated. Once it's compensated, its attenuation can be measured using a sine-wave signal. Connect a sine-wave generator to the probe and supply a 1000 Hz signal. With the probe connected to the scope's vertical input, establish a peak-to-peak display of any convenient number of divisions. Without adjusting the generator or scope controls, disconnect the probe from the generator and supply a direct input from the generator to the scope's **vertical input** terminals. To calculate the probe attenuation, divide the number of vertical divisions obtained during direct waveform display by the number obtained when the probe was used.

CHECKING VERTICAL RESPONSE

The frequency response of the vertical amplifiers in a service scope is critical. Any attenuation of high-frequency input signals results in distorted waveform presentations which can lead to erroneous interpretations. Periodic checks of the vertical amplifiers will prevent these mistakes. If a stable square-wave generator is available, its output may be used for this. Connect the output of the generator to the scope and observe the resultant display. Adjust the frequency compensation network in the vertical attenuator circuit as required. A sawtooth voltage also can be used.

The EICO 1460 oscilloscope has compensation adjustments for the calibrated **vertical attenuator**. Individual trimmer capacitors are provided for each of the (10, 100, 1000) positions on the control. Adjustment of the frequency response of the attenuator is made by correctly setting each trimmer. Referring back to Fig. 6-4, a jumper is connected between the **sawtooth** and **vertical input** terminals and the **horizontal**

selector set to +. The **sweep selector** is then set to the **100-1K** position and the **sweep vernier** to **100**. The **vertical gain** control goes to the **50** position, while the **horizontal gain** control is set to **40**. With the **vertical attenuator** set at **10** (either ac or dc), the display is centered using the positioning controls. If the attenuator is not correctly compensated, the trace will appear as shown in B and C of Fig. 11-4. The trimmer capacitor for the attenuator position under test is adjusted to remove the hook from the display (Fig. 11-4A). After the attenuator has been compensated at the **10** position, the same procedure is employed, using separate trimmers, for the **100** and **1000** positions.

A much preferred method of making these same adjustments on the EICO 460 is to apply a square wave of approximately 1 kHz to the **vertical input** terminals and adjust each trimmer in turn (with the **vertical attenuator** selector set to the corresponding position) for best possible square-wave reproduction. Internal sweep is employed and the panel controls set for several stationary square waves on the screen. Figure 11-3 shows the results of under- and overcompensation.

For a detailed analysis of the vertical amplifiers, a sweep frequency generator should be used. Since most service scopes should have an approximate bandwidth of 0 to 5 MHz, the sweep generator should have a provision for an output that is swept from 0 to at least 5 MHz. The output of the sweep generator is applied to the input of the vertical amplifiers while a second scope is used to observe the resultant waveform. Using a low-capacitance probe, the entire waveform envelope will be displayed. A correct display is one similar to that of waveform 1 in Fig. 11-5A. The envelope is fairly flat out to the upper limits of frequency response; then, as high-frequency attenuation starts to occur, the envelope starts to close. Waveform 2 results when there is a steady increase in attenuation as frequency increases.

Individual stages of the vertical amplifier may be checked using the same procedure. Keep in mind, however, that the stages may be stagger tuned to achieve the desired overall bandwidth. If that is the case, then the first stage may have a considerable high frequency gain and the following stages a high-frequency attenuation.

Trouble with the high-frequency response in the amplifiers can usually be traced to peaking coils or lead dress within the amplifier. Low-frequency problems are commonly caused by the resistors.

CHECKING THE TRIGGERING CONTROLS

Practically all service-type oscilloscopes have some sort of triggering control. These controls establish the point at

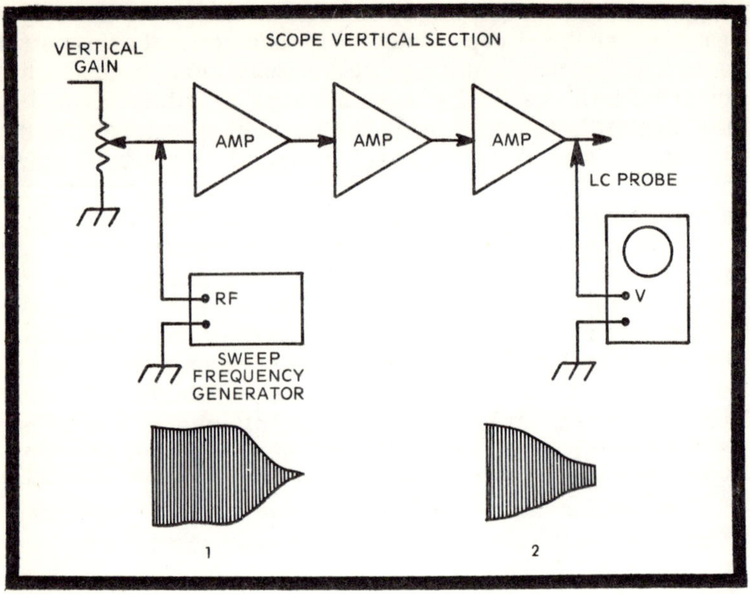

Fig. 11-5. Checking the frequency response of the vertical amplifiers.

which the sweep circuit is triggered. A typical triggered-sweep scope has two such controls. One, a **polarity** control, selects which polarity of the trigger pulse will be used to start the sweep circuit. The other control is usually a **trigger level** control that sets the amplitude at which the trigger signal starts the trace. If an **automatic** provision is made, the scope will trigger on the average level of the waveform.

To check the operation of these controls, almost any waveform may be used, as long as it has easily determined polarities and distinguishable levels. Here, a sine wave will be used.

The **level** control may be checked with the **trigger polarity** control set to either the **positive** or **negative** position. In Fig. 11-6, the **polarity** control is set to the **positive** position. With the **level** control near its zero position, the displayed waveform will show nearly the entire leading edge, as seen in waveform A. As the **level** control is moved to about the center position, the display is shifted in such a manner that the waveform starts nearer its positive crest, as in B. A continued increase of the **level** control setting causes the waveform to start at an even higher level, as shown in C.

When the **level** control is moved to its maximum position, the displayed waveform will start at its most positive peak, as

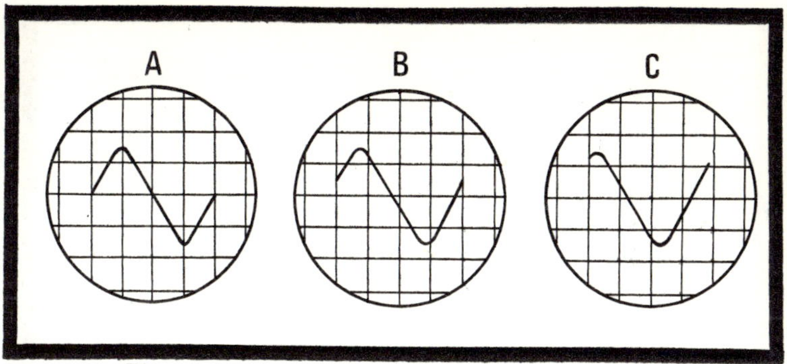

Fig. 11-6. The effects of the trigger level control.

shown in Fig. 11-7A. With the level control at this position, operation of the **polarity** control is easily observed by switching to the **negative** position. Now the waveform should appear 180 degrees out of phase with that previously displayed (Fig. 11-7B). Actually there is no change in polarity of the waveform. The sweep is started at the negative peak of the signal instead of the positive peak.

THE BALANCE AND LINEARITY CONTROLS

The purpose of a **balance** control in either the vertical or horizontal amplifier is to prevent any shift in the zero reference point of the displayed waveform when the position of the **gain** control is changed. For example, increasing the **vertical gain** control should cause an increase in the total height of the display, but the zero baseline should remain constant at all settings of the control. On the other hand, the **balance** control corrects for any change in the size of the displayed waveform caused by a change in the setting of the

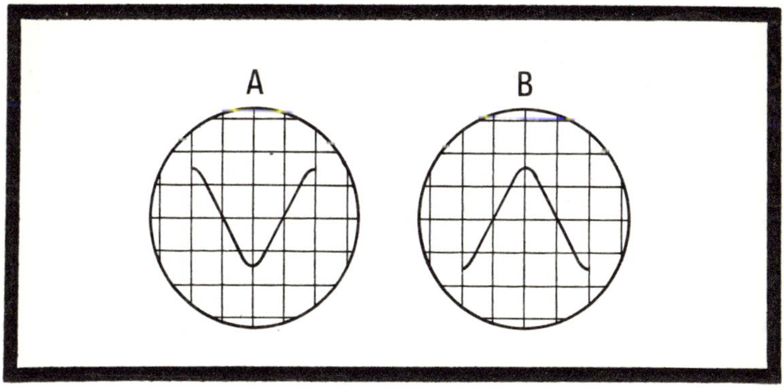

Fig. 11-7. The effects of the trigger polarity control.

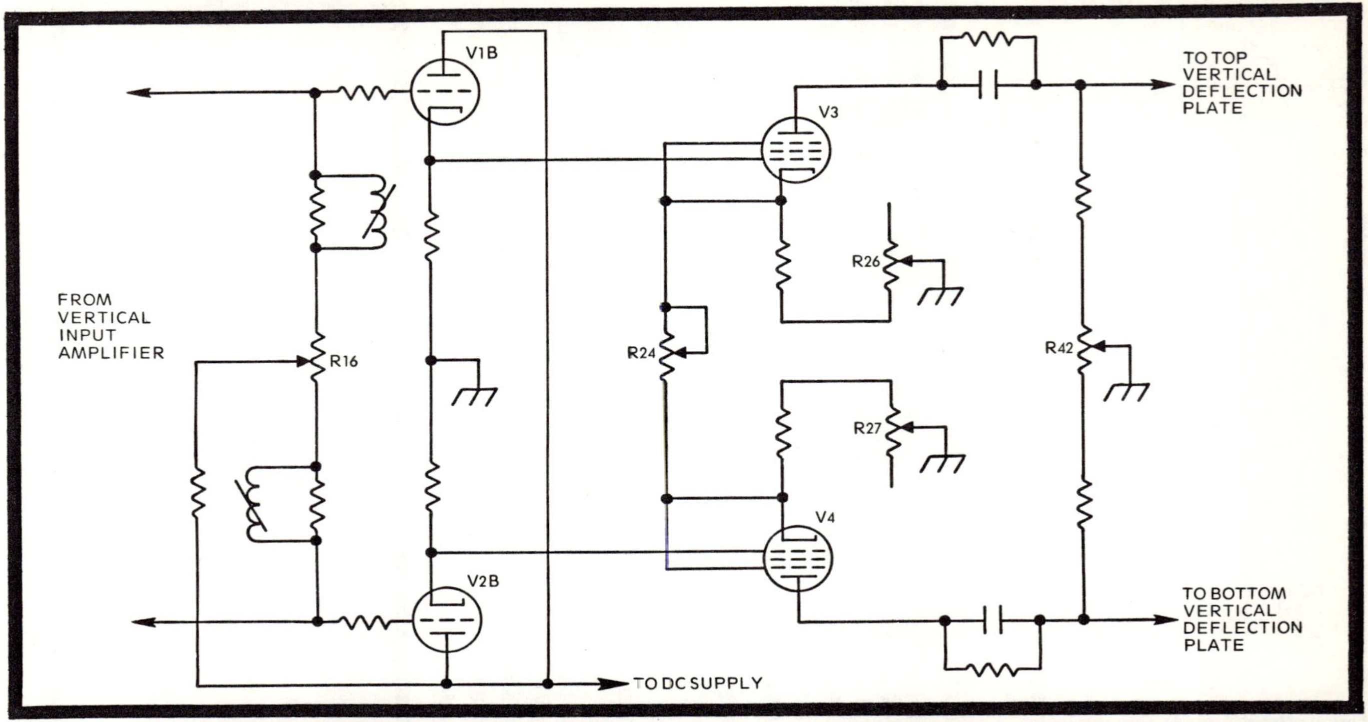

Fig. 11-8. Simplified diagram of the vertical controls. (Courtesy EICO Electronic Instrument Co., Inc.)

positioning control. These controls all function together to achieve a balanced linear display. A simplified schematic of the controls of the EICO 460 appears in Fig. 11-8.

Adjustment begins with R16, the **dc balance** control. With the **vertical input** terminals shorted and the **vertical attenuator** set to the **ac 1000** position, set the **vertical gain** control, R24, for minimum gain. This insures that no signal reaches the deflection plates, and the trace is positioned entirely by the dc voltages in the vertical amplifiers. Adjust R42, the **vertical position** control, for a dead center trace. Now set R24 to maximum. If the amplifiers are balanced, there will be no shift of the trace on the crt. If the trace shifts, adjust R16 until the trace returns to the center of the screen. Return the **gain control** to minimum, observing any movement of the trace and using R16 to bring the trace back to zero. This operation is repeated until there is no vertical shift of the trace when R24 is rotated from maximum to minimum.

After a correct dc balance is obtained, the **linearity controls**, R26 and R27, are adjusted using an EVM. These adjustments are used to set the conduction levels of V3 and V4 to the same level. Then, after sufficient time for V3 and V4 to stabilize at their new levels, the dc balance procedure is repeated.

Once these adjustments have been completed, movement of the **vertical gain** control should not affect the zero baseline of the display, and movement of the **vertical position** control should not change the height of the display.

ASTIGMATISM

In an oscilloscope, the roundness of the electron beam is adjusted by the **astigmatism** control. How beam shape affects the appearance of the trace is illustrated in Fig. 11-9.

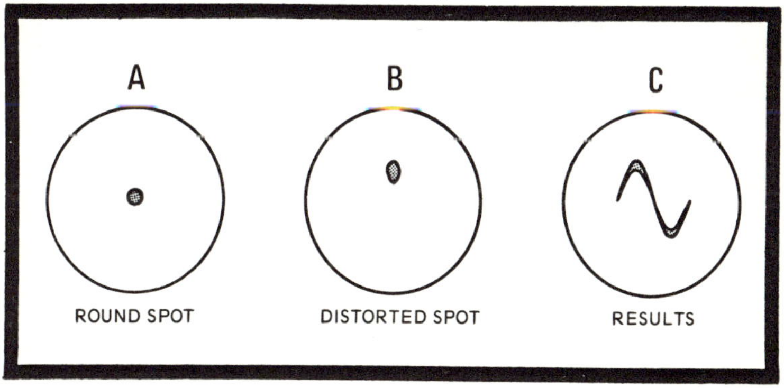

Fig. 11-9. The effects of astigmatism.

Assuming that a perfect beam is produced by the electron gun in Fig. 11-9A, the beam will appear as a round spot at the center of the crt screen when the deflection voltages are zero. As the beam is deflected, as shown in B, the appearance of the spot is distorted due to the angle at which it strikes the surface of the screen. This can be compared to shining a flashlight at your feet and then gradually moving the spot of light across the floor away from your body. The beam of light will take on an elongated shape. As an out-of-round beam is deflected about the screen, the trace broadens, resulting in a distorted display (Fig. 11-9C).

The **astigmatism** control is generally located in the crt bias circuits, and interacts with the **focus** control to determine the sharpness of the trace.

Glossary

Astigmatism Control
The oscilloscope control that determines the uniformity of the trace width.

Bandwidth
The range of frequencies between the two frequencies at which current drops to 70 percent of its value at resonance. In oscilloscopes, the range of frequencies over which the specified amplifier is capable of operation with a defined level of attenuation.

Calibrated attenuator
A signal-dropping device usually contained in the vertical input section of a service oscilloscope to provide a means of reducing the size of the displayed waveform by a specific ratio.

Current probe
A test equipment probe that samples the signal under observation by induction.

Dc-wideband oscilloscope
A term used to describe an oscilloscope that is capable of displaying waveforms that range in frequency from dc to several megahertz, typically 4.5 MHz or more.

Deflection factor
A ratio expressing the amount of beam deflection obtained for a given level of signal input; a measure of the oscilloscope's sensitivity.

Demodulator probe
A probe used to remove an AM signal from its rf carrier.

Dual beam
A term used to describe an oscilloscope in which two electron beams are formed.

Dual channel
A term used to describe an oscilloscope in which two separate vertical input channels exist.

Dual trace

A term describing the capability of a dual-beam or dual-trace oscilloscope to simultaneously display two separate traces.

EVM

Electronic voltmeter; a general classification of voltmeters utilizing internal electronic circuits; a VTVM, TVM, or FETVM.

Expanded scale

A method of displaying a waveform with an oscilloscope wherein only a small portion of the total waveform is displayed, causing the displayed portion to appear larger.

External sweep

A method of applying a sweep voltage from a source outside the oscilloscope to the horizontal deflection plates.

External sync

A sync signal received by an oscilloscope from an external source.

FETVM

Field-effect transistor voltmeter; an electronic voltmeter employing field-effect transistors in the electronic circuits.

Frequency range (sweep)

The operating range of the horizontal sweep circuits of an oscilloscope, describing the limits of the sweep repetition rate.

Graticule

A transparent overlay on the viewing screen of an oscilloscope that provides a scale with which to judge the displayed waveform. May be calibrated in inches or centimeters.

High-voltage probe

A test lead used to extend the limits of the voltage scales of a voltmeter by dropping a large portion of the applied voltage across a high resistance contained in the probe.

Input capacitance

The amount of shunting capacitance to ground offered by the input circuits of a test instrument. The higher the input capacitance, the more seriously a high-frequency signal will be distorted.

Input impedance

The combined effect of input capacitance and input resistance. The input impedance of a test instrument will determine the amount of circuit loading caused by its connection to a circuit.

Input resistance
The resistance seen by the circuit under test when a test instrument is connected.

Intensity
The brightness of the trace on the screen of a crt; also the control that adjusts brightness.

Intensity modulation
Varying the brightness of the trace of the crt by varying the intensity of the beam with an external signal.

Internal sync
The act of synchronizing the horizontal sweep circuit with a portion of the input waveform.

Isolation probe
A test lead containing a resistive element to increase the input impedance of a meter and thereby help prevent circuit loading.

Linearity
The quality of maintaining a constant relationship. Used in referring to the fidelity with which the vertical and horizontal aspects of a waveform are reproduced on an oscilloscope.

Low-capacitance probe
An oscilloscope test lead that lowers the shunting capacitance offered to the circuit under test.

Low-power ohms mode
A mode of operation of an EVM wherein the test voltage applied to the circuit by the meter is lowered to a point at which solid-state devices will not be forward biased.

Ohms adjust control
An adjustment used to set the meter pointer to a reference when measuring resistance with an EVM.

Overshoot
Distortion of a waveform displayed on an oscilloscope, caused by the electron beam overtravel on the leading or trailing edge of a pulse.

Phasing control
A control in the sync circuits of some oscilloscopes that enables the operator to adjust the phase of the sync signal when it is derived from the powerline voltage.

Protection diode
A solid-state diode used to protect the movement of a meter circuit from excessive voltage.

Recurrent sweep

The horizontal sweep of an oscilloscope when the repetitive triggering of the sweep is achieved by the action of circuits within the oscilloscope. (Compare the **triggered sweep.**)

Rise time

The elapsed time of a waveform between the time it reaches 10 percent of the maximum height of the leading edge and the time it reaches 90 percent of that height.

Sensitivity (meter)

A measure of the input impedance of a meter. Determines the amount of circuit loading caused by the meter.

Sensitivity (scope)

A ratio of applied signal voltage to beam deflection.

Triggered sweep

The horizontal sweep of an oscilloscope when the repetition rate of the sweep and its associated triggering are controlled by the signal applied to the vertical input.

TVM

Transistor voltmeter; an EVM employing transistors in the electronic circuits.

Undistorted deflection

The maximum amount of deflection that can be achieved in an oscilloscope before the waveform becomes distorted.

Vectorscope

A type of oscilloscope in which the demodulator output signals of a color television are compared as vectors.

Vernier

A fine adjustment. For example, the sweep speed controls of an oscilloscope may consist of a coarse setting and a vernier (fine) control.

Z-axis

The intensity modulation axis.

Zero adjust

The control used for balancing the amplifier circuit in an EVM.

Index

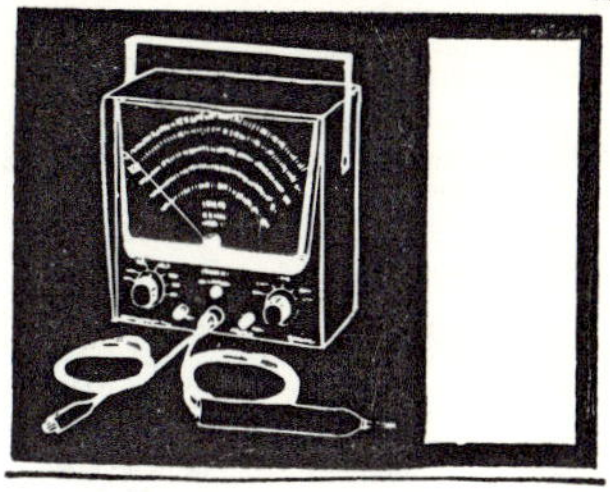